Popular Art Deco Lighting
Shades of the Past

Herb Millman and John Dwyer

Schiffer Publishing Ltd
4880 Lower Valley Road, Atglen, PA 19310 USA

Dedication

To our three guardian angels: Fran Dwyer, Abe Millman, and Richard Perez

Library of Congress Cataloging-in-Publication Data

Millman, Herb.
Popular art deco lighting : shades of the past / by Herb Millman and John Dwyer.
p. cm.
ISBN: 0-7643-2043-2 (hardcover)
1. Electric lamps--Collectors and collecting--Catalogs. 2. Electric lamps--History--20th century--Catalogs. 3. Lighting, Architectural and decorative--Collectors and collecting--Catalogs. 4. Lighting, Architectural and decorative--History--20th century--Catalogs. 5. Decoration and ornament--Art deco--Collectors and collecting--Catalogs. I. Dwyer, John. II. Title.
NK8438.M55 2004
621.32'09'041075--dc22
2004002556

Designed by Bonnie M. Hensley
Cover design by Bruce Waters
Type set in Broadway BT/Humanist 521 BT

ISBN: 0-7643-2043-2
Printed in China
1 2 3 4

Published by Schiffer Publishing Ltd.
4880 Lower Valley Road
Atglen, PA 19310
Phone: (610) 593-1777; Fax: (610) 593-2002
E-mail: Info@schifferbooks.com

For the largest selection of fine reference books on this and related subjects, please visit our web site at **www.schifferbooks.com**
We are always looking for people to write books on new and related subjects. If you have an idea for a book please contact us at the above address.

This book may be purchased from the publisher.
Include $3.95 for shipping.
Please try your bookstore first.
You may write for a free catalog.

In Europe, Schiffer books are distributed by
Bushwood Books
6 Marksbury Ave.
Kew Gardens
Surrey TW9 4JF England
Phone: 44 (0) 20 8392-8585; Fax: 44 (0) 20 8392-9876
E-mail: info@bushwoodbooks.co.uk
Free postage in the U.K., Europe; air mail at cost.

Contents

Acknowledgments

Without the support of our families, friends, clients, and the community of New Hope, we could not have come this far. Thanks to so many, we have expanded our retail showroom space, written two books, and are now taping a weekly television series.

It is nice to hear from those who visit our store that they envy our coming to work each day to a place that is filled with so many beautiful things. Many who visit our shop find themselves awed by the beauty of our lights. Then get hooked. They continuously return to see what treasures we bring into our shop. There are countless individuals that we have to thank, not only for their continued support but some for their help in finding us some of our treasures. Bob and Dee Dee Corey, Matt and Tricia Dromboski of Lamptiques, Louise Gulmantovicz and Jeffrey Skinner of The Roosters Coup, Joe Mattis of The Black Swan, Hildegard Mervis of Antique & Unique Lighting, Ron Lillich of Past Ranch, Vanessa Strougo of Deco Etc., and—last but not least—Randy Wicker, formerly of Uplift Lighting. Randy was not only our mentor for many years but also taught us and guided us as both a teacher and a very special friend.

We must also thank several of our friends and clients, many of whom allowed us to photograph a number of pieces that they purchased from us. Tony, Jennifer, Patrick & Vivian Artur, Ed Collins & Jose Ortiz, Michael & Lisa Hlatke, John Donelan, Laura Goddard, Joe Knox & Rick Sweeney, Michael Kuric & Jack Tyler, Peter Prorok & Ramon Robledo, Bill Schudder, Les & Jan Sdorow, Dr. & Mrs. Jesse Silverman, Eugene Smith & Ed Norski, Dr. Rey Velasco & Frank DeLuca, Henry & Lori Ward, Richard Werner & Paul Barunas, Axel Wieschenberg, and last but not least Andrew Zourides, who has been a dear friend for over twenty years.

When it comes to a major support system, family rates as number one. We have to take this opportunity to thank our families for their tremendous support over the years. Jean Dwyer & Robert Burton, Kathleen King, Stan & Merle Millman, Scott, Danielle & Hunter Millman, James & Felicia Liepa, Joseph, Robert & Alison Goldman, and all of John's brothers & sisters too numerous to mention.

We must also thank a very special lady, Penny Sandor. Thanks to her confidence in us, we were able to expand our retail space and receive a wider exposure in the community.

We must also take this opportunity to thank Peter and Nancy Schiffer. Without their confidence, patience, and guidance, we would never have had the opportunity to create this second book on art deco lighting.

Lastly, we would like to dedicate this book to our three Guardian Angels: Fran Dwyer, Abe Millman, and Richard Perez.

Preface

So many of us take things for granted. Many years ago, prior to the invention of the electric light bulb, candles, kerosene and gas lights were the source of interior light. With electric lighting, we have choices. Today we have lighting from incandescent to fluorescent to halogen. We have soft white and bright white and color tinted light bulbs. Today lighting plays a very important part in commercial settings as well as how one will decorate his or her home. Due to a large market, many companies are reproducing styles of lighting of years gone by.

However, there is nothing like the real thing. Original art deco lighting has an enormous amount of detail throughout the framework and glass. Whether we are looking at chandeliers, wall sconces or table lamps, the detail is quite obvious.

Today, many commercial venues such as movie theatres, banks, and even shopping malls are installing art deco reproduction lighting. "Why?" one might ask. The answer is, the art deco period has made a major resurgence because of our nostalgia for a modern but simpler time when there was both hope and idealism. You can walk into lobbies of new office buildings, theatres or restaurants and you will see the signs of yesteryear with touches of the art deco period.

We, however, do not believe in reproductions. This book contains photographs of original art deco works of art. And yes, to us original art deco lighting is a work of art. This book's photographs include intricate detail of lighting that was constructed to last. Today, to make lighting objects with such intricate detail, not to mention the materials that were used in the art deco period, the price would be very dear. Even though white metal was used to manufacture many of the pieces, bronze cast iron, brass aluminum and other alloys were also used to construct frames and bases. The interior electrical workings were also of a higher quality. Porcelain sockets and cloth wrapped wires were used and many of the connections were soldered onto the sockets. Today many of the fittings are constructed of light aluminum and different types of durable plastics.

Lights, Camera, Action. We enjoy watching old movies. These old movies had the original lighting accents on the set. One of our favorite films is *Vogues of 1938*, which was filmed in 1937. To us the acting is secondary and the sets are primary. Even in later films such as *Chinatown* or *Mommie Dearest*, you know that they must have searched high and low for the original deco pieces. Obviously, when something could not be found, it was reproduced to appear to be original. Once again, one might ask, why? Perhaps it is because the art deco period had so much style in every sense of the word. It was theatrical. It was worldly. It made a statement. And in some films, it was as important as any character.

Introduction

Every book has an introduction to introduce the reader to the authors, to the topic, and to give a background for the book. So here is ours. We wrote our last book, *Art Deco Lighting*, in 2001. That book was a result of our publishers coming to our store and being impressed with our large inventory of twentieth century lighting. We were surprised and delighted when the Schiffers asked us to write a book on this subject. And now, with the success of our first book and our better than ever inventory, they want us to do a second book.

The chandeliers, sconces, table lamps and torchieres we sell have not just put bread on our table, but have given us great pride and joy. (It is rewarding when you are of service and have helped people light and decorate their homes. The home is a refuge and its style is a deliberate reflection of its owners. In our lighting as well as our furniture and decorating advice, we try to give a comfortable sense of history–both personal and general.) We know by selling our beautiful lighting, we have affected people's lives, their everyday lives. The elegance and charm of "the first modern age" is a style that appeals to most of us in the twenty-first century. In this, our second book, we hope to look at more lighting of the period and more closely examine the literal radiance of art deco lighting.

A note on reproductions: There are some companies reproducing lighting. Though the majority of the companies that distribute the lighting are American based, the labor is often done in Indonesia and China. These pieces vary in quality. Some of them are good, but of those that are true reproductions, we have yet to see a reproduction that matches the quality of the original. This is both in the materials used and the aesthetic quality of the piece.

We do want to distinguish for the reader a difference between a reproduction and a new piece. We often hear customers use these words interchangeably. They are not the same. A reproduction is a piece that is a copy. It is a look-alike. There is an original that preceded it. There are many new pieces that are being done in the Art Deco Style. They have no precedent.

No Facsimiles

(No Facts! Silly Millies!)

And now, a little story. Yesterday, a woman came into our store with a figural lamp that she wanted to sell. She actually looked like the actress who is currently starring on Broadway in *Thoroughly Modern Millie*, Sutton Foster. Her hair was bobbed with bangs. She looked like a 1920s girl, thin with big eyes. She compared her lamp to a period radio lamp in the store that was marked six hundred dollars and she wanted to sell it to us for at least that amount.

We told her that her lamp was most likely made in the 1980s. It was of a woman coming out of a tree trunk with a large etched glass panel fanning out behind her shoulders. It stood about two feet tall and was cream colored with an ivorene face. She insisted it was from the 1940s at least. She had bought it from a dealer that said he believed it to be old but did not really have any information on it.

Her story then changed from "it was totally original" to "it had been rewired." She had to say that because neither the cord nor socket was replaced by anything remotely suggesting age. But there was a problem to the rewiring, if it were a rewiring of an old piece. The problem was that the wiring that replaced the piece was inappropriate to an old piece. On an old piece that gets rewired, the owner generally tries to preserve the aesthetic of the old piece with a dark brown silk cord and the correct socket. The plastic, see-through cord and the white-cardboard tapered socket sang out "New Piece!" In turning the lamp upside down to check for markings, there was a small label that said "Made in China" on its bottom. Thoroughly Modern Millie immediately scratched off the small label upon our mentioning it to her.

She then said that, aesthetically, it was nicer than some older pieces. This was true. But then she said what to us is the ultimate wrongheaded notion. The ultimate sin! "I buy a piece for how it looks! I don't care about its age!" Millie yapped. And yap she did. For five minutes, she told us how age doesn't matter. "Yap! Yap! Blah! Blah! Yap! Yap!" she pleaded. She really wanted to sell that lamp but her argument was unnerving. Herb went out for a cigarette. John escorted her to the door, politely saying we weren't interested at the moment.

Age does matter. A respect for history is important. Original pieces, whether lighting or furniture, carry a history with them. There is a spiritual quality in an original lamp, in an antique, that cannot be replicated in Indonesia. The comfort of a home is not just a visual aesthetic. It is a comfortable sense of permanence and stability. In a dot-com world, where we surround ourselves with dot-com furniture that seems more appropriate for office buildings or airports than what one would have in a home, we can lose a sense of what is truly important. The past is important, both in general and personally. If we do not believe what people did in the past matters, how can we believe what we do in the present matters. The home is a better place when those "shades of the past" are surrounding you. It reminds you of both family and values, in the true non-political sense of the words. Antiques and heirlooms are a connection to the human family. It acknowledges the contributions of the past and infers the significance of the present.

So take that, you silly Millies!

As to the monetary value of antiques, and in this book's particular area lighting, value tends to vary according to circumstance. Aunt Phoebe gives you an heirloom. It's free! You pick up something at a yard sale. The value of a good high-end piece, however, is set by a knowledgeable seller and the market.

In our book we price according to what we would expect to pay in a retail store. If you are buying from a store in a small town in the Midwest, expect the lower end of the price range. If you buy in midtown New York, expect the higher end of the range. Deco is a style more highly prized in urban centers and, due to supply and demand as well as overhead, prices are generally higher in the big city.

Besides retail stores, you can buy, of course, at yard sales, flea markets, antique shows, the internet, low-middle end auctions, and high end auctions (Sotheby's or Christie's). If the piece is for your personal use, I would stay away from the internet for your main lighting, unless you know the seller (i.e., you know of the store "John Doe Lighting", saw the piece at the store, and now are buying it over the net or by phone). Lighting is important enough in your home for you to actually see the piece before buying it. If you are a dealer, you can afford to get something that was not exactly what you thought. You can always resell it. The homeowner, however, is stuck if the piece is photogenic but not "exactly right" in person.

When you're older, you realize that lighting is important. It can create romance, mystery, and excitement. And with older antique lighting, the romance often is more intense. We hope that, unlike Millie, you value period art deco lighting. We hope that you enjoy the book. We have been blessed to write it.

Accent & Figural Radio Lamps

As we indicated in our first book, ***Art Deco Lighting***, accent and figural lamps were mass-produced from the 1920s through the 1940s. The compositions of these lamps were white metal (spelter), cast iron, brass, and bronze. Many of these lamps were designed with a tremendous amount of intricate detail. Various finishes were applied to the lamps to highlight these details, while others were hand-painted to accentuate facial expressions and design features.

A popular motif was the female form, often nude or partially nude. The deco style meant that the figure was in motion: dancing, diving, lunging. Birds, exotic animals, dogs, and last, but not least, man were also subject matters for these lamps. But the female nude was, as Arthur Von Frankenburg (founder of Frankart–maker of period accent lamps, book ends, and ashtrays) said, the highest form of nature. And whether clothed or not, a beautiful woman was the most popular subject matter for figural lamps.

A number of these lamps were originally designed to rest on newel posts at the bottom of stairwells. These were considered an elegant accent in prestigious homes.

Others were meant to sit on the early floor model radios. In the 1930s, as the family would gather around the radio after dinner, the lights would be dimmed to allow the imagination to soar. While listening to tales of "The Green Hornet" or to Jack Benny or Burns and Allen, the only light in the room would be the glow from the radio lamp atop the radio.

Many of the shades and globes were manufactured in Czechoslovakia. They were comprised of cased glass and were multi-colored in design. American manufactured globes and shades were mostly made in a solid color with intricate designs throughout the glass. Most popular are the American made Brain glass and Crackle glass globes.

Most recently, some globes and lamps are being reproduced. The reproductions lack the weight, intricate detail, and character of the figural lamps of the past.

1930s art deco radio lamp. Composition: white metal with bronze wash. Globe is pale yellow with black bands. 10" W x 10.5" H x 4" D. $695-800.

Late 1920s art deco dual figural accent lamp with two ashtrays and match/cigarette holder. Composition: white metal with original dark green patina and brass accents. Crackle globe has been replaced. 11.25" H x 12" W x 4" D. $1100-1250.

Early 1920s art deco figural lamp. Composition: White metal with original green patina. Original glass back plate, .25". 9.5" H x 5.5" W x 5" D. $825-950. A very rare find.

1930s accent figural lamp. Composition: cast base, nickel-plated center post with female nude riding horse. Light cylinders are original white milk glass with original metal caps. 11" H x 13.5" W x 5" D. $875-1000.

Early 1930s art deco silhouette figural lamp. Composition: brass with original paper background. 8.125" D. 9" H. Very unusual. $865-975.

Late 1920s-early 1930s art deco radio lamp. Composition: alabaster with intricate detail. Female nude white metal with pewter wash, holding clear ball. 11" H x 13.5" W x 6.75" D. $1265-1500.

Late 1920s French art deco accent lamp. Composition: base marble and onyx; chrome-plated figure with period globe. 5" H x 9.25" W x 3.75" D. $715-840.

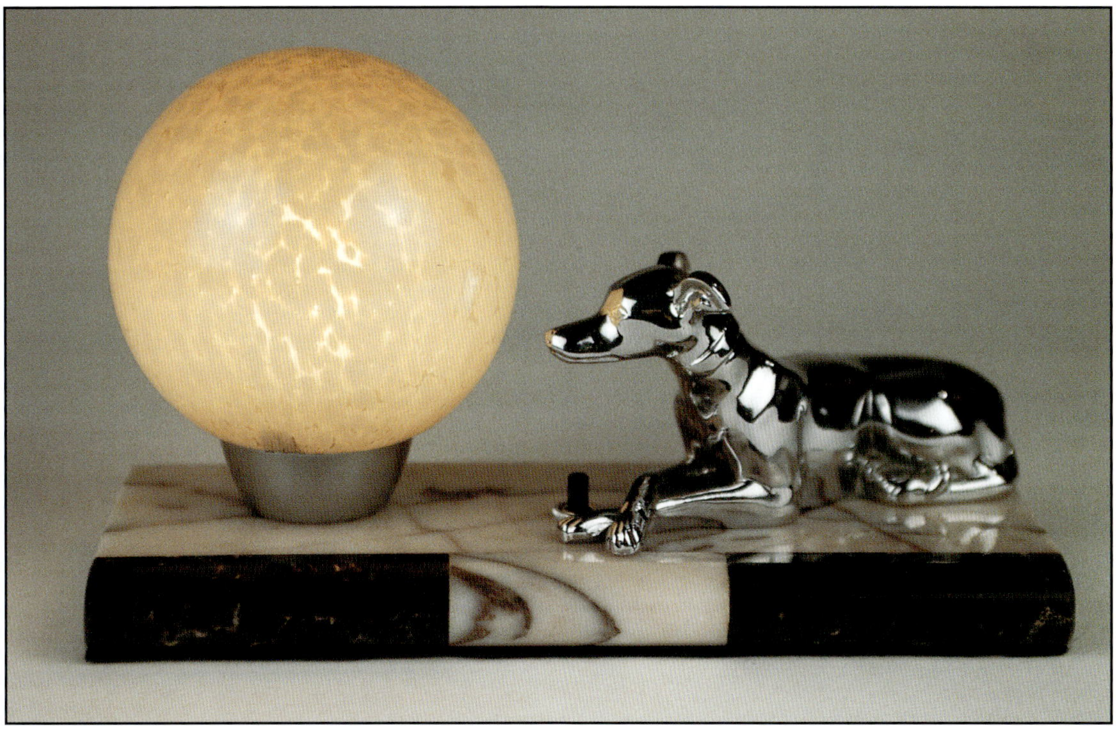

1930s German art deco boudoir lamp with Czechoslovakian shade. 12.5" H x 2.75". $425-490.

1930s art deco figural lamp boasts three female nudes. Composition: white metal with bronze wash. Intricate details throughout the base. Oversized seeded glass globe of the period. 14.25" H x 5.75" W. $1125-1350.

1930s German art deco radio lamp. Alabaster base, white metal female figural nude, copper-plated. 14" H x 4.5" W. $960-1100.

1930s art deco "Nuart" radio lamp. Original pearlized finish with period beehive crackle globe. Composition: base, cast; figure, white metal. 7.25" H x 8.75" W x 5.5" D. $1095-1400.

1930s art deco dual figural nude "Nuart" radio lamp. Period globe boasting female nude. Globe is a combination of pressed glass and amber satin glass. A very rare find. Composition: white metal with gold/ bronze wash. 17" H x 5.5" W x 4" D (base). $1725-2000.

1930s art deco boudoir lamp. Composition: white milk glass shade and base. Female imprinted on milk glass. 10.5" H x 3.25" square. $495-575.

1930s art deco white female nude boudoir lamp. Composition: ceramic and milk glass. Shade has been replaced, 12.5" H x 5.25" W, $425-500.

1930s art deco radio lamp. Composition: white metal, bronze wash; period green crackle glass globe. 8.5" H x 9.5" W. $585-725.

1930s dual figural radio lamp. Composition: white metal with bronze wash, highly glazed. Czechoslovakian globe. 8" H x 10" W x 3.5" D. $595-695.

1930s art deco boudoir lamp. Composition: white frosted satin glass with aluminum sailboat mirrored on glass shade in black. 9.75" H x 6" W x 4" D. $295-350.

1930s art deco dual figural female nude radio lamp. Composition: white metal with a black wash. Period globe, white satin frosted with black bands throughout. 10.5" H x 11.25" x 5" D. $985-1100.

Late 1920s German art deco dual figural radio lamp. Composition: alabaster with white metal, brass-plated female figures. A very rare find. 8.75" H x 15.5" W x 3.75" D. $1390-1650.

1930s art deco sailboat lamp. Composition: aluminum with black bands, white satin glass globe. 12" H x 5.5" D (base). $360-420.

1930s French art deco. Composition: brass with period globe. 6.5" H x 8.5" W x 2.75" D. $495-600.

1930s art deco silhouette lamp. Composition: white metal. Glass has been replaced. 11" H x 12" W x 4" D. $465-525.

1930s art deco boudoir lamp. Composition: white satin glass. Lamp base boasts female nude. 9.75" H x 4.5" W x 4" D. $495-575.

1930s art deco dual figural lamp. Czechoslovakian end-of-day glass globe. Composition: white metal with matt black finish. 15" H x 5.25" W x 4.25" D. $765-850.

Early 1930s art deco radio lamp. Lamp base boasts three female nude figures with intricate details throughout. Composition: white metal with bronze wash; globe fitter is copper housing end-of-day Czechoslovakian globe. 15" H x 7.35" W x 4.5" D. $795-895.

1930s art deco female nude figural lamp with ashtray. Composition: spelter with bronze wash, period brain-glass amber globe. 11" H x 7.5" W x 5" D. $495-575.

Early 1930s French art deco mantel clock with battery operated light at base of clock. Composition: base, marble; mermaid, white metal with a bronze wash. Clock is a wind-up. Base houses original Bakelite switch for battery light. 5.5" H x 12.25" W x 3" D. $795-875. (A very rare find)

1930s art deco female nude figural lamp, "Nuart." Composition: white metal with matt black finish, satin glass ribbed cone shade. 8.25" H x 8" W x 5" D. $765-900.

Early 1930s art deco figural nude accent lamp. Composition: alabaster. 10.75" H x 6" W x 5" D. $1195-1400.

1930s art deco chrome-plated figural lamp. Original satin glass cylindrical shade with metal and chrome cap. 12.5" H x 6" D (base). $610-700.

Early 1930s art deco figural lamp. Composition: base, Akro Agate; figure and lamp holder, white metal with original patina; hands and face, ivorine. Shade has been replaced. 18" H x 5.5" W (base) x 4.25" D (base). $1195-1400.

Late 1920s-early 1930s art deco table lamp. Composition: white satin cased frosted glass. Shade boasts three female nudes resting on the moon surrounded by stars and half moons. This "Tiffen" shade is a very rare find. 13" H x 7" D (shade), 4.5" D (base). $995-1200.

1930s art deco radio lamp. Composition: cast iron and white metal with original patina; hands and face are ivorine. 9" H x 10" W x 4.75". $985-1100.

Early 1930s art deco table lamp. Composition: alabaster, Bakelite, ivorine, and bronze. Arabian costume. Cold painting. This alabaster base is the actual lamp. 16" H x 5.5" D (base). $1300-1750.

1930s art deco radio lamp. Very strong Egyptian influence. Composition: white metal with bronze wash; glass back plate has been replaced. 16.75" H x 5.75" W (base) x 5.75" D. $925-1100.

1930s French art deco perfume lamp. Composition: ceramic. Signed "Argilar Paris." 6" H x 8" W x 3.25" D. $1600-2200.

1930s art deco figural radio lamp. Lamp base boasts two horses in gallop. Composition: alabaster base and globe; white metal horses with a pewter wash. 8.25" H x 14" W x 5.5" D. $685-800.

1930s art deco "Nuart" table lamp. Female figure holds ashtray. Composition: white metal with bronze wash; shade has been replaced. 15" H x 9" W x 5.25" D. $865-950.

Late1920s art deco dual figural lamp. Very strong Egyptian influence. Composition: wrought iron and white metal; mica inserts within the pyramid. 17" H x 15" W x 4.5" D. $1365-1550

1930s art deco "Nuart" radio lamp. Composition: white metal with original green patina; period globe, white frosted satin glass. 8.25" H x 8.75" W x 5.5" D. $895-1100.

Late 1920s art deco male nude figural lamp. Composition: base and globe, alabaster; male figures, white metal with bronze and pewter wash. A very rare find. 7.25" H x 14" W x 5" D. $1250-1600.

1930s art deco accent figural radio lamp, signed "Frankart." Composition: white metal, original green patina; original glass panel, .25-inches thick. 11" H x 6.5" W x 5" D. $1295-1700.

1930s art deco radio lamp. Composition: alabaster base, period amber globe; figure white metal with bronze wash. 10.5" H x 10" W x 4.75" D. $745-850.

1930s art deco dual figural nude radio lamp, signed "Frankart." Composition: white metal with original black wash; original amber crackle glass globe. Very hard to find. 9" H x 13.5" W x 5.5" D. $2750-3400.

1939 World's Fair table lamp of the Trylon and Perisphere was handmade in 1940. Composition: wood. 15.5" H x 14" W x 8.25" D. $385-495.

1930s French art deco figural lamp, signed "Sabino No. 4903 France." Composition: molded pressed glass staircase; white metal figures, hand-painted with a bronze wash. 19" H x 8.75" W x 8.5" D. $995-1600.

1920s French art deco figural lamp. Harlequin influenced by Ballet Russe. Composition: silver plate over bronze; period beaded shade. Signed "A. Leclerc." 24.5" H x 5.5" W x 3.25" D. $2200-2800.

Art deco-style male nude figural lamp; not dated, however it is period. Composition: plaster; globe has been replaced. 22.5" H x 18.5" W x 5.25" D. $975-1200.

1920s French art deco table lamp. Female with two Borzoi dog figures. Composition: white metal with pewter and gold tone wash; base is marble and onyx. A most exceptional and rare lamp. (French foundry stamp located on the rear of the figure.) 16.75" H x 23.75" W x 4.25" D. $3600-4500.

Early 1930s art deco table lamps. Very unusual. Composition: brass and chrome with Bakelite switches and white satin frosted shades. 13.5" H x 5" D (base). $685-825 pair.

1920s art nouveau table lamp. Composition: brass and white metal with bronze wash. 21" H x 9.5" D (base). $985-1325.

1930s art deco female nude figural radio lamp. Composition: white metal with pewter wash, end-of-day Czechoslovakian globe. 13.75" H x 11" W x 6" D. $825-995.

1930s art deco dual female nude figural lamp. "Nuart." Composition: white metal and cast iron, Czechoslovakian globe. 9" H x 14.5" W x 6.5" D. $985-1265.

1930s art deco female nude figural radio lamp. Composition: white metal and cast, frosted satin green glass globe. 8" H x 9" W x 5" D. $745-935.

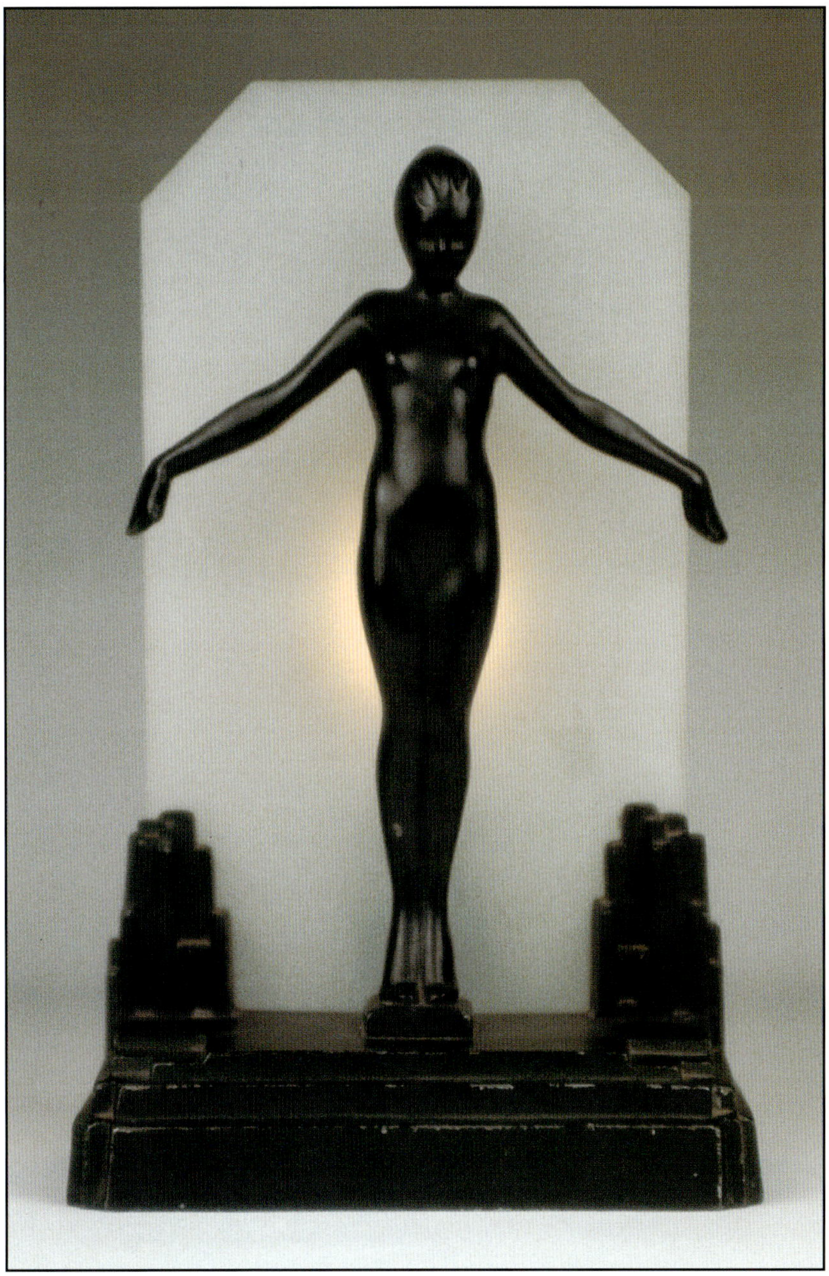

1930s art deco "Frankart" female nude accent lamp. Composition: white metal and cast iron. Original frosted glass panel, .25-inches thick. 10.5" H x 6.5" W x 5" D. $1295-1700.

1930s art deco female figural radio lamp. Composition: white metal with bronze wash and Czechoslovakian globe. 6.5" H x 8.75" W x 6" D. $675-825.

1930s art deco female figural accent lamp. Composition: white metal with pewter wash, cobalt blue glass has been replaced. 10.25" H x 6.75" W x 3.75" D. $475-650.

1930s art deco female nude figural lamp. Composition: white metal, pewter wash, multicolored antique satin Czechoslovakian globe. 10.75" H x 5.25" W x 3.25" D. $675-850.

Early 1930s art deco dual figural female nude radio lamp with mercury glass globe. When lit it appears orange. Composition: white metal. 8.5" H x 11" W x 8" D. $985-1200.

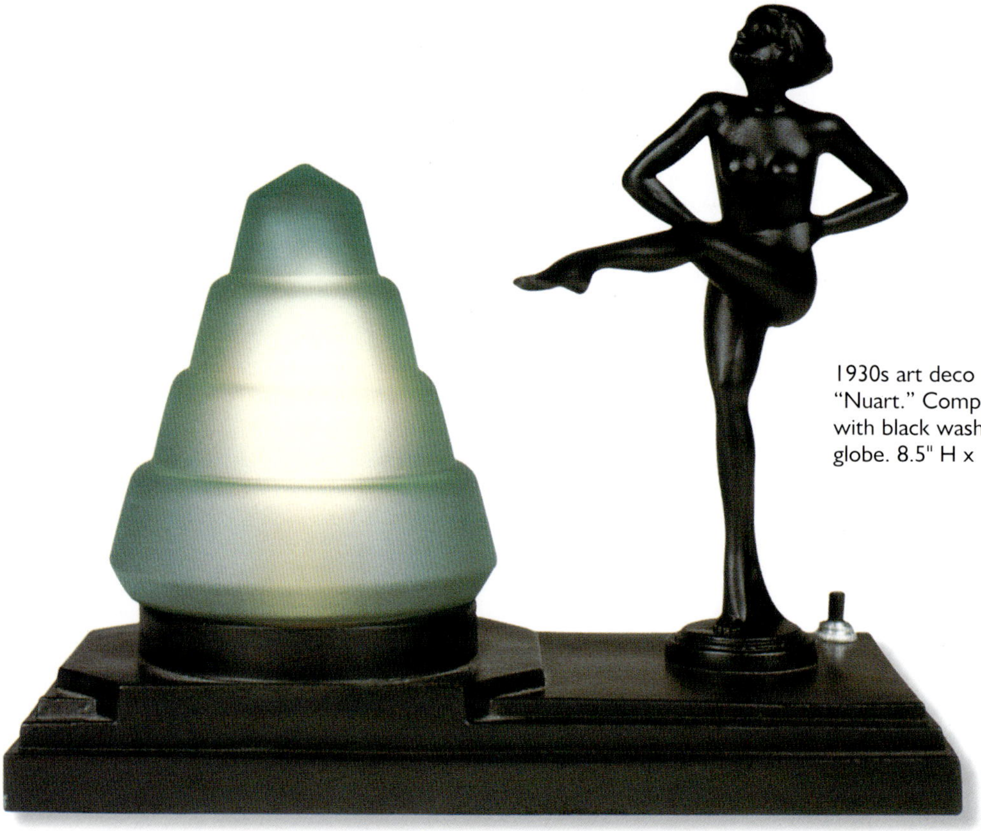

1930s art deco female nude figural lamp, "Nuart." Composition: white metal and cast with black wash. Emerald green satin glass globe. 8.5" H x 10.25" W x 5.5" D. $985-1200.

Early 1930s art deco dual figural radio lamp. Composition: white metal, chrome-plated, seeded glass globe. 11.5" H x 6" W x 5" D. $625-750.

Early 1930s art deco female nude accent lamp. Composition: spelter with pewter wash, globe has been replaced. 8" H x 4.5" W x 5.25" D. $410-525.

1930s art deco "Nuart" female nude table lamp, clutching ashtray. Composition: white metal with pearlized wash. 14.25" H x 6" W x 7.5" D. $685-825.

1920s French art deco female nude figural lamp. Composition: white metal on cast base. Period globe. 10" H x 15" W x 5" D. $995-1350

1920s art deco female nude figural lamp. Composition: stepped marble base, white metal figure with pewter wash. Globe has been replaced. 25.5" H x 7.5" W x 14" D. $1250-1600.

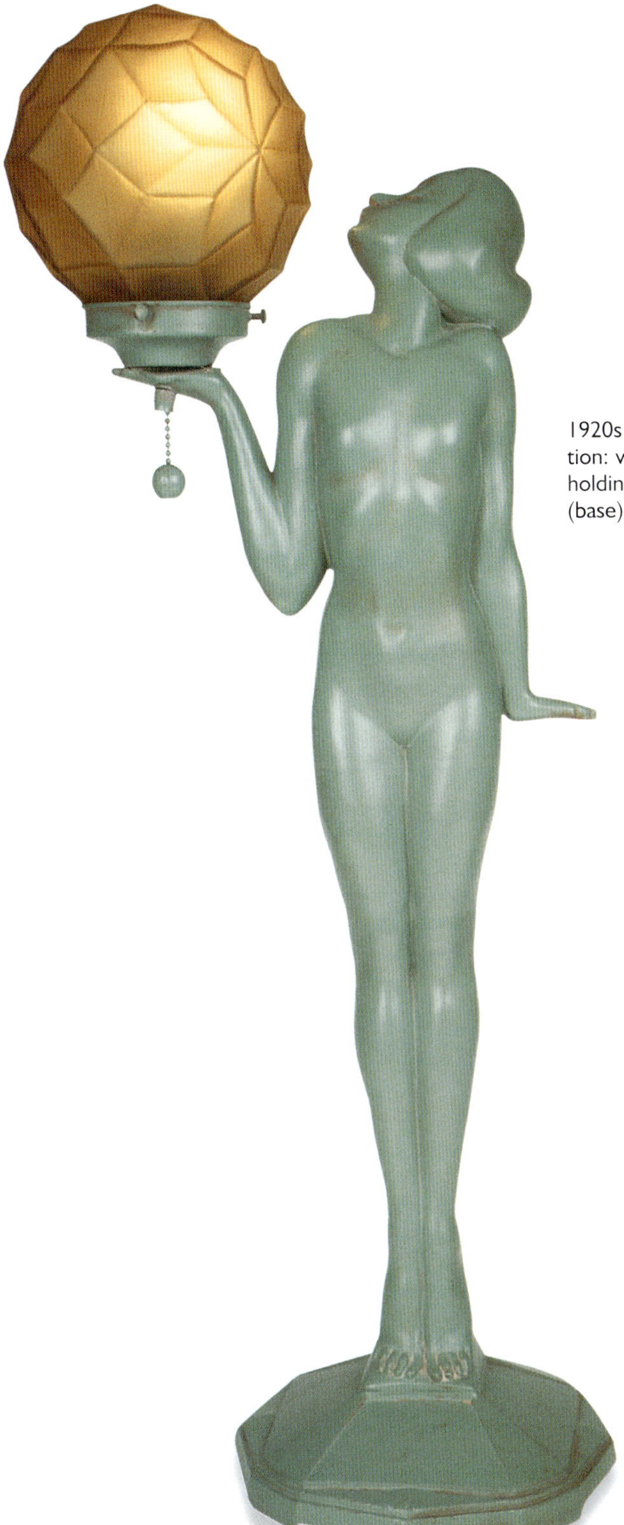

1920s art deco female nude figural lamp. Composition: white metal and cast, original green patina, holding amber geometric globe. 29.25" H x 8.5" D (base). $1295-1800.

Early 1930s art deco female nude figural lamp, "Nuart," with unusual emerald green crackle shade. Composition: white metal with bronze wash. 8.75" H x 9.75" W x 4.25" D. $995-1250.

1930s art deco accent lamp. Reverse painting on white frosted satin glass. 11.5" H. $425-500.

1930s art deco accent figural lamps. Two elephants flank the base of each lamp with their trunks raised. The globes are geometric in design, all of satin frosted glass with a combination of cased and pressed glass. "A Very Rare Find." $600-700 pair.

Late 1920s/early 1930s art deco accent lamp. Reverse painted male figure on original frosted glass panel. Frame and base, white metal. $600-725.

1930s art deco accent lamps. Composition: chrome plated, which make for "A Very Rare Find." 18.5" H. $695-850.

1930s art deco accent lamps. Composition: white milk glass shades with chrome-plated bases (a very sleek design). $650-800 pair.

Early 1930s art deco smoking stand. Composition: white metal with cast base. Center shaft boasts a sculpture of the "Statue Of Liberty" with a light bulb as the torch. Never before seen, "A Very Rare Find." $950-1200.

1930s art deco accent lamps. Composition: white metal, black glass bases, and white frosted horizontal ribbed shades. Two gazelles flank eaach base. Overall dimensions: 8" W. x 8.5" H. x 5.5" deep. $750-900 pair.

Boudoir Lamps

From the 1920s through the 1940s, boudoir lamps adorned vanities, mantles, and buffets. They were originally designed for the bedroom vanity, but found their way into other rooms in the home.

A majority of boudoir lamp shades were phallic in design. Many were manufactured of pressed glass as well as cased and frosted glass. The shades were designed with various ribbed details such as etching, ribbing, and smooth multi-colored satin glass finishes.

Boudoir lamp bases were constructed from a wide variety of materials. Most common were pressed glass bases. However, bases were also made of chrome, white metal, and bakelite, as well as silver and cobalt blue mirror.

1930s art deco boudoir lamp. Composition: brass base, cased glass shade with blue and frosted accents. 9.5" H x 5" D (base). $115-140.

1930s art deco boudoir lamp. Composition: chrome base with blue satin glass shade. 11" H x 3.25" D (base). $110-145.

1930s art deco boudoir lamp. Composition: chrome base and cap, white frosted ribbed glass shade. 10" H x 4" D (base). $110-165.

1930s art deco boudoir lamp. Composition: white frosted satin glass and aluminum. 9" H x 5.5" x 3.25" D. $325-395.

1930s art deco boudoir lamps. Composition: jadite green cased glass with black enamel and chrome accents. 11.25" H x 4" W x 4" D. $750-945 pair.

1930s art deco boudoir lamps. Composition: green satin frosted glass ribbed cylinders with chrome bases and caps. 12" H x 6.5" D (base). Very hard to find. $595-750.

Early 1930s art deco boudoir lamps. Composition: white metal with pewter and black accents, original frosted glass panels. 12" H x 5" W. $695-795 pair.

1930s art deco boudoir lamps. Composition: copper base and amber pressed glass shades. 11" H x 3.75" D (base). $365-425 pair.

1930s art deco boudoir lamps. Composition: Bakelite with brass accents, original parchment paper shades. 14.75" H x 3.75" W x 3.75" D. $295-375 pair.

1930s art deco boudoir lamps. Composition: pressed glass with Bakelite accents. 12" H x 4" D (base). $295-385 pair.

1930s art deco boudoir lamps. Composition: mirrored bases with chrome accents, pressed glass shades. 10.75" H x 4" square (base). $295-385 pair.

1930s art deco boudoir lamps. Composition: aluminum bases, pull chain lights, white frosted shades with silver relief on glass. 11.25" H x 4.5" D (base). $295-385 pair.

1930s art deco boudoir lamps. Composition: pressed glass shades with pressed glass bases. 11.5" H x 5" D (base). $235-290 pair.

1930s art deco boudoir lamps. Composition: white ribbed milk glass shades with chrome bases and original top caps. 12.75" H x 4.75" D (base). $495-595 pair.

1930s art deco boudoir lamps. Composition: cobalt blue bases with pressed glass dome and chrome shaft. Shades have been replaced. 14.25" H x 4" W x 4" D. $315-425 pair.

1930s art deco boudoir lamps. Composition: chrome with banded white satin frosted glass shades. 11" H x 4.5" D (base). $395-485.

1930s art deco boudoir lamps. Elegant female sweep on each lamp. Composition: shades are frosted pressed glass. Bases and figures are frosted satin glass with original cream-colored sockets and Bakelite switches. 9" H. $600-750 pair.

1930s art deco boudoir lamps. Each lamp base boasts a female nude figure. Composition: molded frosted satin glass. $650-800 pair.

Table & Desk Lamps

From the 1920s through the 1940s, art deco table and desk lamps were constructed from a variety of materials. They ranged from white metal, alabaster, cast iron, brass, and bronze, to ceramic. Many of the white metal and cast iron lamps were either chrome plated or had akro-agate accents. Chrome was considered an elegant, machine age look. It became quite popular on both table and desk lamps. A number of table lamp bases were designed with carved intricate detail. Again, just like radio and accent lamps, a popular motif was the female form in motion.

Over the years, fabric and parchment shades were replaced due to wear and cleaning. A majority of the fabric shades in this chapter have been replaced with new ones. We have carefully selected shades that would be appropriate and conform to the lamp in both color and design.

Late 1930s art deco boudoir lamp. Composition: Akro Agate base, brass accents, and ceramic center column. Shade has been replaced. 13" x 4" square base. $195-250.

1920s art deco accent lamp. Composition: brass with original patina, mica inserts throughout. 15.75" H x 5.25" W x 5.25" D. $385-500.

1930s art deco table lamp. Composition: ceramic with brass base. Shade has been replaced. 14" H x 3.5" D (base). $115-145.

1930s art deco table lamp. Composition: brushed aluminum. 11" H x 4" D (base), 7" D (shade). $315-400.

1930s art deco chrome table lamp. Shade has been replaced. 16" H x 5.25" D (base). $395-425.

1930s art deco chrome desk lamp, unusual design. 12" H x 5" D (base) x 8.25" W overall. $725-850.

1930s art deco alabaster table lamp. Very unusual coloration. 14" H x 5" D (base). $565-710.

Brass table lamp with strong art deco influence. Composition: brass, with replaced finial 13" H x 7" D (base). $89-125.

1930s art deco table lamp. Composition: aluminum, machine age in design. Shade has been replaced. 17.25" H x 4" D (base) $485-600.

Early 1930s art deco table lamp. Composition: alabaster. Unusual design. 12" H x 9" D (base). $1100-1475.

1930s art deco alabaster table lamp. Unusual in design. 10" H x 4.5" D (base). $725-900.

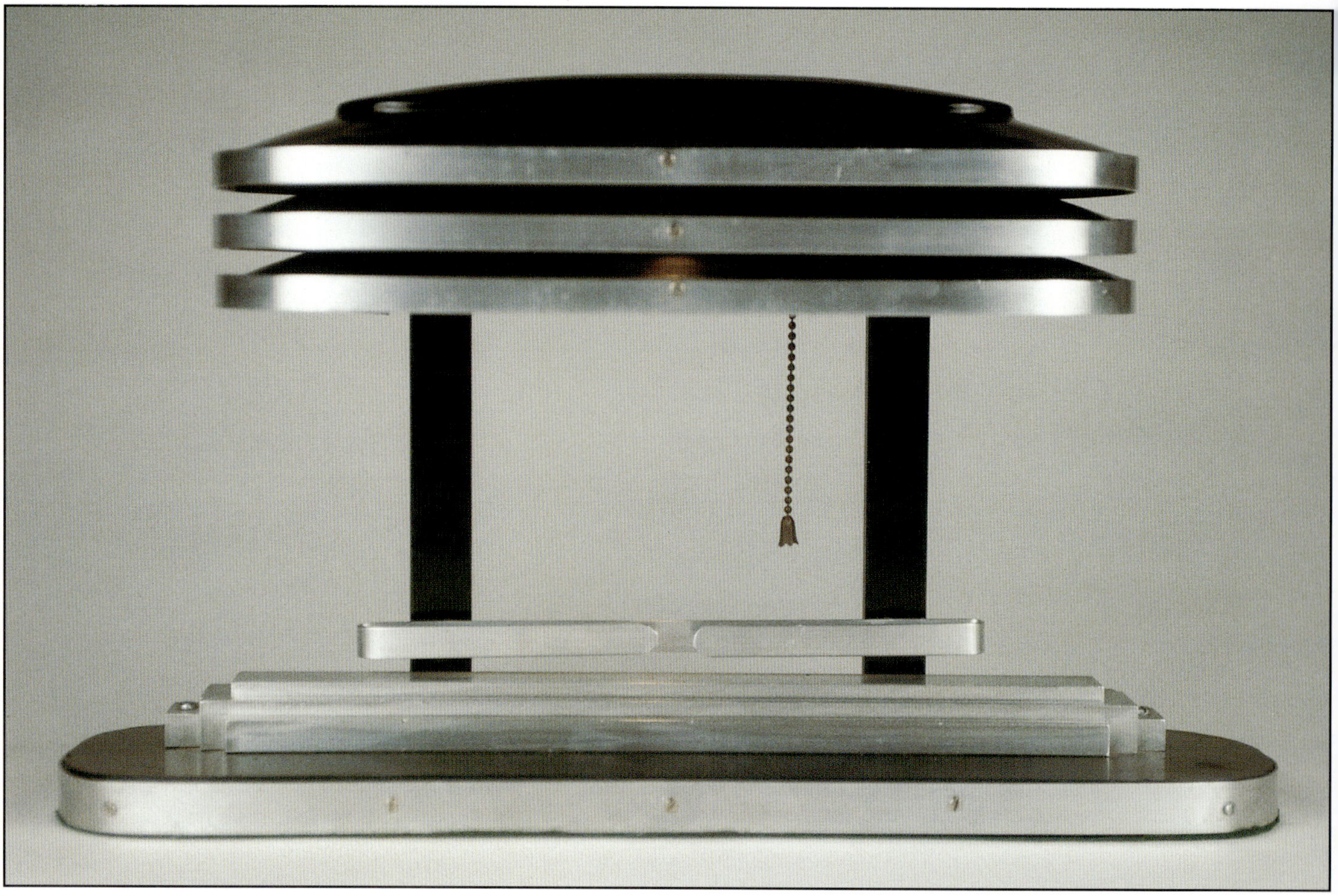

1930s art deco desk lamp. Composition: chrome, aluminum, wood, enamel. Unusual lines. 11.5" H x 19" W x 4.5" D. $995-1175.

1930s art deco table lamp. Composition: chrome and black enamel; shades have been replaced. 15.75" H x 4.75" D (base). $375-450.

1930s art deco desk lamp. Composition: chrome with enameled center shaft. 14" H x 6" D (base). $495-585.

1930s art deco desk lamp. Composition: chrome and aluminum. 20" H x 6.75" D (base). $395-565.

1930s art deco brass desk lamp. Shade and finial have been replaced. 15" H x 4.25" W x 3" D. $225-295.

1930s art deco female nude table lamp. Composition: brass base and center shaft, ceramic figure, glaze painted. Shade has been replaced, but finial is original. 23.5" H x 8.25" W x 3.75" D. $965-1200.

1930s art deco chrome gooseneck desk lamp. 19" H x 7" D (base). $415-525.

Early 1930s art deco table lamp. Composition: ceramic with brass base and shaft. Shade and finial has been replaced. 29" H x 7" D (base). $715-900.

1930s art deco table lamp. Very intricate design. Composition: brass and Akro Agate. Shade has been replaced; finial is original. 30" H x 7.75" square. $585-725.

Early 1930s art deco table lamp. Very industrial in design. Composition: brass and enamel. Shade and finial have been replaced. 24.5" H x 7" D (base). $725-875.

Early 1930s art deco table lamp. Composition: chrome with cast iron base. Shade and finial have been replaced. 26" H x 7" D (base). $499-625.

1930s art deco table lamp. Composition: alabaster with intricate detail. Shade and finial have been replaced. 28.5" H x 6" D (base). $595-685.

1930s art deco banker's lamp. Composition: center shaft, emerald green glass; base, brass with original metal shade and finial. 21" H x 5" D (base). $495-600.

1930s art deco table lamp. Composition: brass, both the finial and the shade have been replaced. 19.5" H x 5.05" square base. $395-450.

1930s art deco table lamp. Composition: chrome and enameled aluminum. 21.5" H x 9" D (base). $465-550

Late 1920s French art deco table lamp. Composition: chrome-plated. Aztec influence with intricate detail. Shade and finial have been replaced. 28" H x 9.25" D (base). $1095-1400.

Early 1920s art deco table lamp. Strong Egyptian influence, intricate detail over pyramid base. Composition: silver gilding over bronze with the original patina and leaded glass shade. 29" H x 8" W x 8" D. $3500-4000.

Early 1930s art deco table lamp. Composition: Akro Agate base with brass accents; original fabric shade with glass beads. 34" H x 6" D (base). $950-1600.

1930s art deco table lamps. Composition: cased glass, custard in color. 9.25" H x 4.25" D (base). $395-465 pair.

1930s art deco chrome accent table lamps. 12.5" H x 5.25" D (base). $635-725 pair.

1930s Canadian art deco table lamps. Composition: genuine cut crystal; shades and finial have been replaced. Paper label "Made in Canada." 19.75" H x 7.75" W x 5" D. $565-695 pair.

1930s art deco table lamps. Composition: chrome and glass. Shades have been replaced with paper shades. 15.25" H x 6" D (base). $510-595 pair.

1930s art deco table lamps. Composition: copper and black enamel. Industrial in design. Shades have been replaced. 23" H x 6" W x 3" D. $1195-1400 pair.

Late 1930s art deco Saturn table lamp. Composition: chrome and Bakelite. Shade has been replaced. 15.25" H x 5" D (base). $320-410.

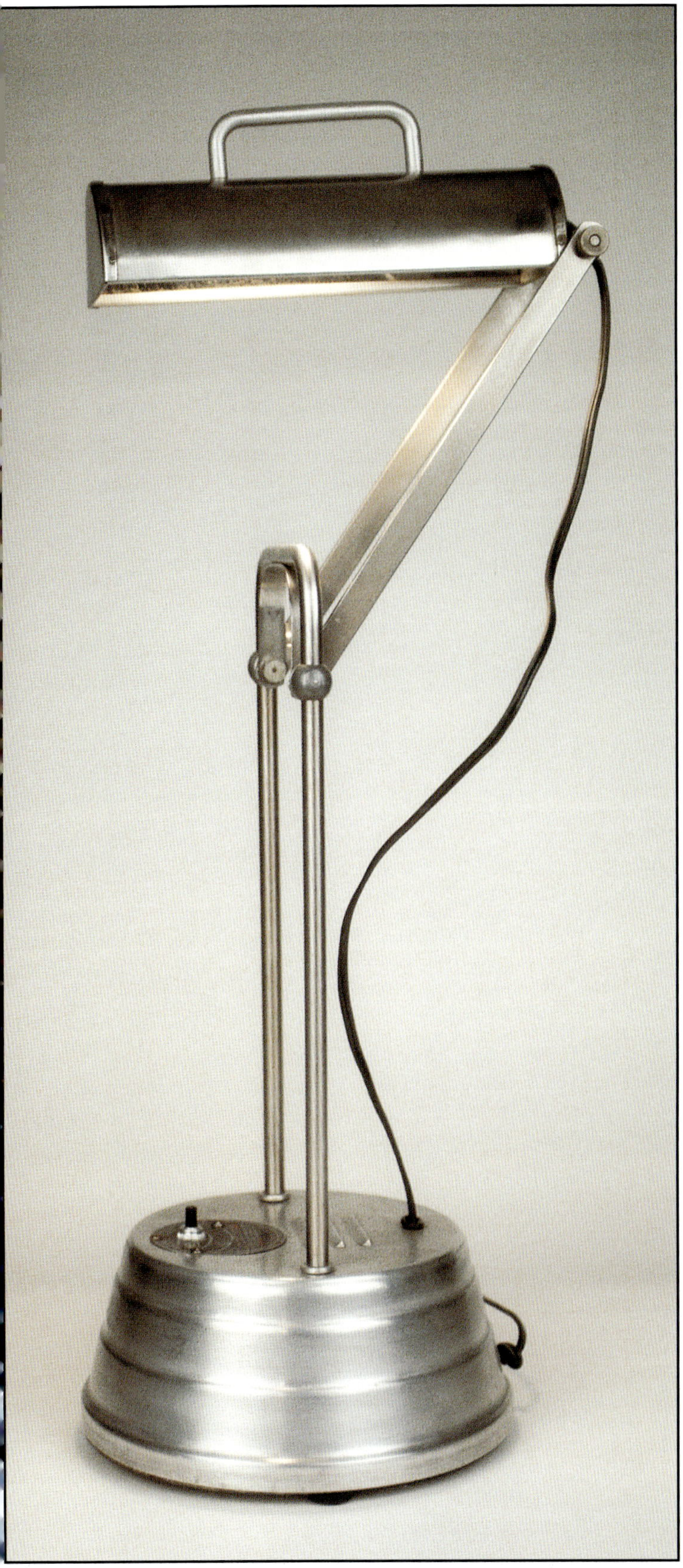

1930s desk lamp converted from heat lamp. Composition: brushed aluminum and chrome. 24" H x 8.75" D (base). $495-650.

1930s art deco "Chase" desk lamp. Composition: solid brass with original Bakelite switch. 11.75" H x 5" D (base) x 7.25" D (shade). $365-425.

1920s art deco female nude figural lamps on alabaster bases. Composition: white metal with pewter wash. Shades and finials have been replaced. 26.5" H x 4.5" D (base). $1400-1750 pair.

1930s art deco table lamps. Composition: copper plated with aluminum interior. 14" H. $1200-1400 pair.

Late 1930s art deco table lamps. Composition: base square cut glass with spun aluminum circular insets. Shades are spun aluminum. 18.5" H. $675-900 pair.

1930s art deco desk lamp, chrome plated. 22.5" H. $650-800.

1930s art deco desk lamp. Composition: spun aluminum with copper accents. 19" H. $550-625.

Overhead & Wall Lighting

This chapter contains photographs of overhead and wall lighting fixtures that were designed and manufactured from the 1920s through the 1940s. Pendant fixtures that were manufactured of milk glass and opalescent glass were very popular in commercial venues, as well as in residences, as were slip shade chandeliers and wall sconces. Many of the fitters used for pendant fixtures were accompanied with matching ceiling canopies. The fitter and canopy were generally constructed of white metal and were chrome plated, while others were made of brass and aluminum.

Many of the manufacturers created slip shade chandeliers and ceiling fixtures with matching wall sconces. Slip shade chandelier frames were designed with intricate detail throughout the frames and were often accompanied with a matching ceiling canopy. The compositions ranged from white metal to brass to bronze and nickel. A number of frames were hand-painted to highlight their detail. A few chandeliers were manufactured with separate on/off switches that allowed the owner to control the lighting directly from the fixture, as well as the wall switch. A majority of the backplates that were manufactured for the wall sconces had on/off switches at the base of the fixture.

The slip shades themselves were manufactured of cased glass and pressed glass. A majority of the shades carried a wide variety of detail throughout the shade, which ranged from a raised geometric design to assorted floral patterns. The shades came in a variety of colors. Most popular were the amber and white frosted slip shades, though a raised satin finish was popular as well.

1930s art deco slip-shade wall sconces. Composition: white metal back plate with strong geometric influence. Opalescent cased glass shades. 9.5" H x 5.5" W x 7" protrusion from wall. $1299-1350 pair.

1930s art deco wall sconces. Composition: white metal with bronze wash. Slip-shades are amber satin cased glass. Strong art deco etching in glass. 10.25" H x 7.25" W x 5.25" protrusion from wall. $1125-1350 pair.

1930s art deco "Moe Bridges" wall sconces. Composition: back plates are white metal with intricate art deco detail and two-tone gold wash. Amber frosted detailed cased glass slip-shades. 11.25" H x 7.5" protrusion from wall. $955-1095 pair

Late 1930s art deco wall sconces. Composition: back plate white metal with four horizontal brass bands. Bubble glass bottom. 7" H x 9" W x 4.25" protrusion from wall. $695-785 pair.

1930s art deco slip-shade wall sconces. Composition: back plate is white metal with bronze wash. Slip-shades are pale green. A very rare find. 12" H x 5.25" W x 5.25" protrusion from wall. $995-1110 pair.

1930s art deco slip-shade wall sconces. Composition: back plates are white metal with bronze wash, manufactured by "Puritan." Slip-shades are amber cased glass with floral design at center. 10.5" H x 5.75" W x 5.5" protrusion from wall. $1065-1125 pair.

1930s art deco ceramic wall sconce with geometric designs throughout. 7.5" H x 4.25" W x 4.25" protrusion from wall. $195-215.

1930s art deco ceramic wall sconce with silver overlay and push switch. 6.75" H x 4.5" W x 5" protrusion from wall. $195-215.

1930s art deco ceramic wall sconces with horizontal bands throughout. Original ceramic pull chain tassels and receptacles. 6.25" H x 4" W x 3.75" protrusion from wall. $495-525.

1930s art deco slip-shade wall sconces. Composition: back plate white metal with gold-tone and bronze wash. Very unusual in design. Slip-shades are cased frosted glass with caramel bands running vertically and horizontally. 8.25" H x 7" W x 4.75" protrusion from wall. $1099-1250 pair.

1930s art deco slip-shade wall sconces. Composition: back plates white metal with bronze wash, geometric in design. Slip-shades are pale yellow cased satin glass with inverted steps. 10.25" H x 6.50" W x 4.25" protrusion from wall. $1095-1185 pair.

1930s art deco slip-shade wall sconces. Composition: back plates bronze with two-tone wash. Intricate art deco detail throughout frames. Slip-shades are custard colored cased glass with caramel vertical design. 10.25" H x 6.25" W x 5.5" protrusion from wall. $1099-1245 pair.

Early 1930s art deco slip-shade wall sconce. Composition: back plate white metal with pewter wash. Slip-shade manufactured by Consolidated Manufacturing Co. White frosted glass, intricate floral design at center of shade. 9" H x 6.75" W x 5.5" protrusion from wall. $495-585.

1930s art deco slip-shade wall sconce. Composition: back plate constructed of white metal with matt gold-tone wash. Slip-shade white frosted cased glass with caramel feather design. 11" H x 5.25" W x 4.25" protrusion from wall. $495-545.

Early1930s art deco slip-shade wall sconce. Composition: back plate constructed of cast iron, pewter wash with brass accents. Slip-shade is cased glass with vertical rib design. 9.5" H x 5.5" W x 4.75" protrusion from wall. $465-525.

Early 1930s art deco wall sconce. Very architectural in design. Composition: white metal with vertical arrow design. Slip-shade is amber frosted glass with intricate art deco throughout. 10.5" H x 5" W x 7.5" protrusion from wall. $545-655.

1920s art deco slip-shade wall sconces. Composition: back plates chrome-plated white metal. Slip-shades are very unusual design in cased white frosted glass with amber accents. 11.5" H x 5.5" W x 4.25" protrusion from wall. $1299-1385 pair.

Top Right: 1930s art deco slip-shade wall sconce. Composition: back plate is white metal with bronze wash. Slip-shade is amber cased glass with geometric details throughout. 11.5" H x 4.75" W x 6.75" protrusion from wall. $485-565.

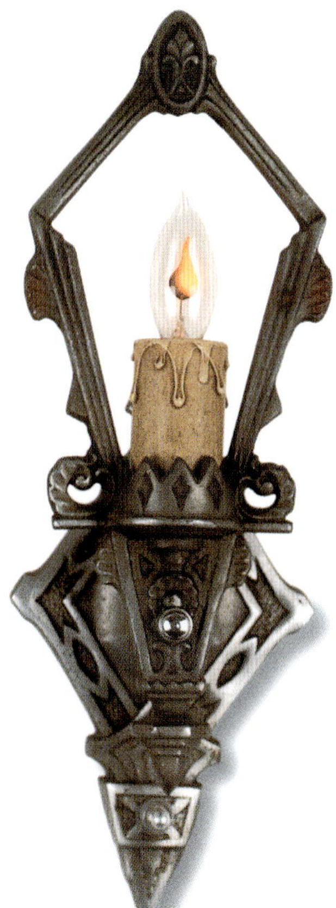

1930s art deco candle wall sconce. Composition: white metal with pewter wash, black and red accents. Architectural in design. 14" H x 4.25" W x 2.25" protrusion from wall. $525-655.

1930s French art deco slip-shade wall sconces. Composition: back plates are silver-plated bronze. Shades are acid-etched cased glass signed "Petitot France." 13.75" H x 12" W x 6.5" protrusion from wall. $2250-2495 pair.

1930s art deco slip-shade wall sconces. Composition: back plates chrome-plated aluminum. 6.25" H x 4.25" W x 7.25" protrusion from wall. $865-925.

1930s art deco slip-shade wall sconces manufactured by "Lincoln." Composition: white metal in gold tone with pewter wash. Amber cased glass with art deco detail throughout. 13" H x 6.5" W x 4.25" protrusion from wall. $985-1075 pair.

1930s art deco slip-shade wall sconce. Composition: back plate is chrome-plated white metal, with brass highlights. Slip-shade is two-tone amber opalescent glass. 9.5" H x 5.75" W x 6" protrusion from wall. $545-610.

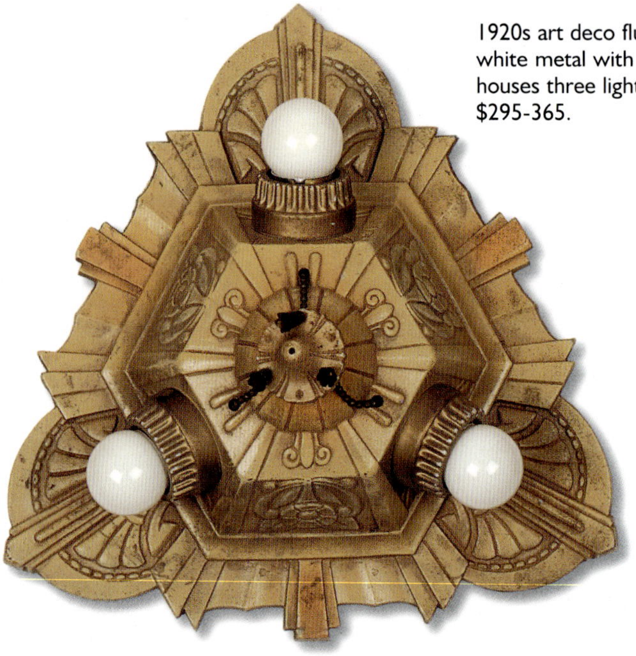

1920s art deco flush mount ceiling fixture. Composition: white metal with original enameled cream-colored paint; houses three lights with center pull chain. 10.5" triangle. $295-365.

1920s art deco five-light flush mount ceiling fixture. Composition: white metal with a bronze wash. It has intricate art deco detail throughout. 17" D. $465-585.

1920s art deco flush mount ceiling fixture houses three lights. Composition: white metal with intricate art deco details throughout. Center finial has been replaced. 11.25" triangle. $265-324.

Early 1930s art deco flush mount ceiling fixture, signed "Markel Electric Products." Composition: white metal with intricate art deco detail throughout, original gold and green paint; shows some wear. 10.75" triangle. $425-535.

1930s art deco flush mount ceiling fixture. Composition: white metal with mirror insets; original cream color finish, center finial has been replaced. Houses two lights. 13.5" W x 6.75" D. $495-535.

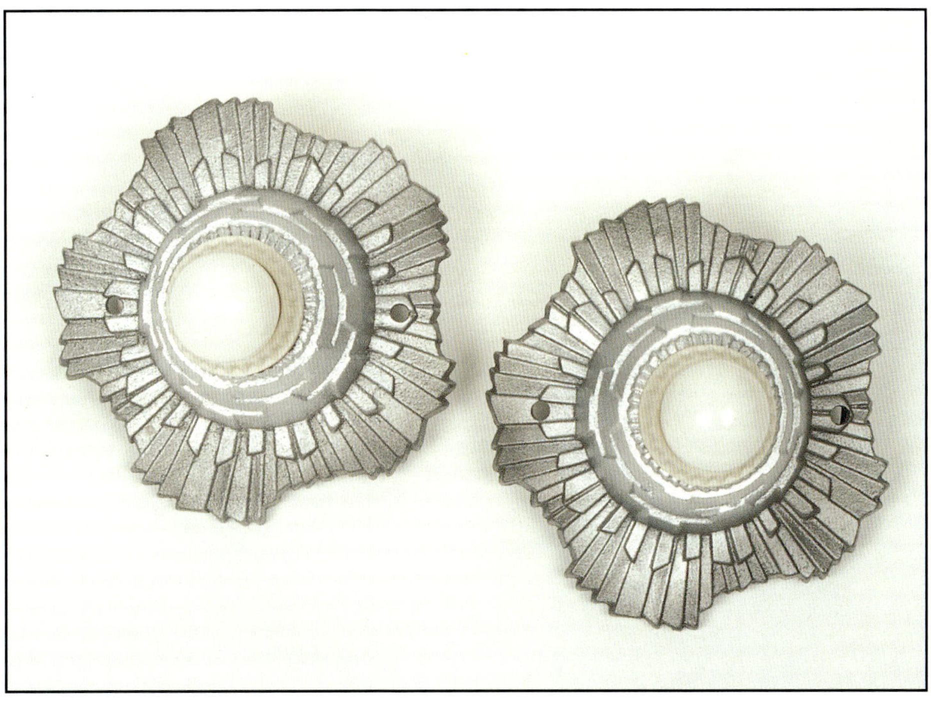

1920s art deco flush mount ceiling fixtures. Composition: cast iron with intricate art deco design throughout. Can also be converted to wall mount fixture. 5.75" D. $235-295 each.

1920s art deco three-light flush mount ceiling fixture. Composition: cast iron with pewter wash. Has intricate art deco detail throughout. 12.75" triangle. $565-775.

1920s art deco slip-shade wall sconce. Composition: white metal with bronze wash. Amber satin cased glass shade. Floral design throughout back plate. 12" H x 5.25" W x 4.75" protrusion from wall. $465-510.

1920s art deco wall sconce signed "Efoufite" and "Durock." Composition: ceramic and chrome. Ceramic is a pale yellow with silver horizontal bands. Cream color cylindrical shade, fluted at top. 9.25" H x 3.5" W x 4" protrusion from wall. $375-425.

1930s art deco wall sconce signed "Efoufite" and "Durock." Composition: ceramic and chrome. Ceramic is a pale yellow with silver horizontal bands. White milk glass cylinder shade. 10.25" H x 3.5" W x 4" protrusion from wall. $375-425.

Early 1930s art deco slip-shade wall sconce. Composition: white metal and brass, lower section with pewter wash. Slip-shade is pale amber cased glass with floral design at center. 11.75" H x 6.75" W x 6.5" protrusion from wall. $435-499.

1930s art deco wall sconce. Composition: white metal with bronze wash; glass panels have been replaced with period glass. 10.75" H x 5.25" W x 6.5" protrusion from wall. $395-425.

1930s art deco slip-shade wall sconce. Composition: back plate is nickel-plated white metal with intricate art deco detail at base. Slip-shade is amber pressed glass. 8.5" H x 5.5" W x 3.75" protrusion from wall. $395-450.

1930s art deco slip-shade wall sconces. Composition: back plates are white metal with pewter wash; slip-shades are a combination of amber and clear horizontal bands constructed of pressed glass. Very unusual in design. 7.5" H x 7" W x 4.5" protrusion from wall. $895-1050 pair.

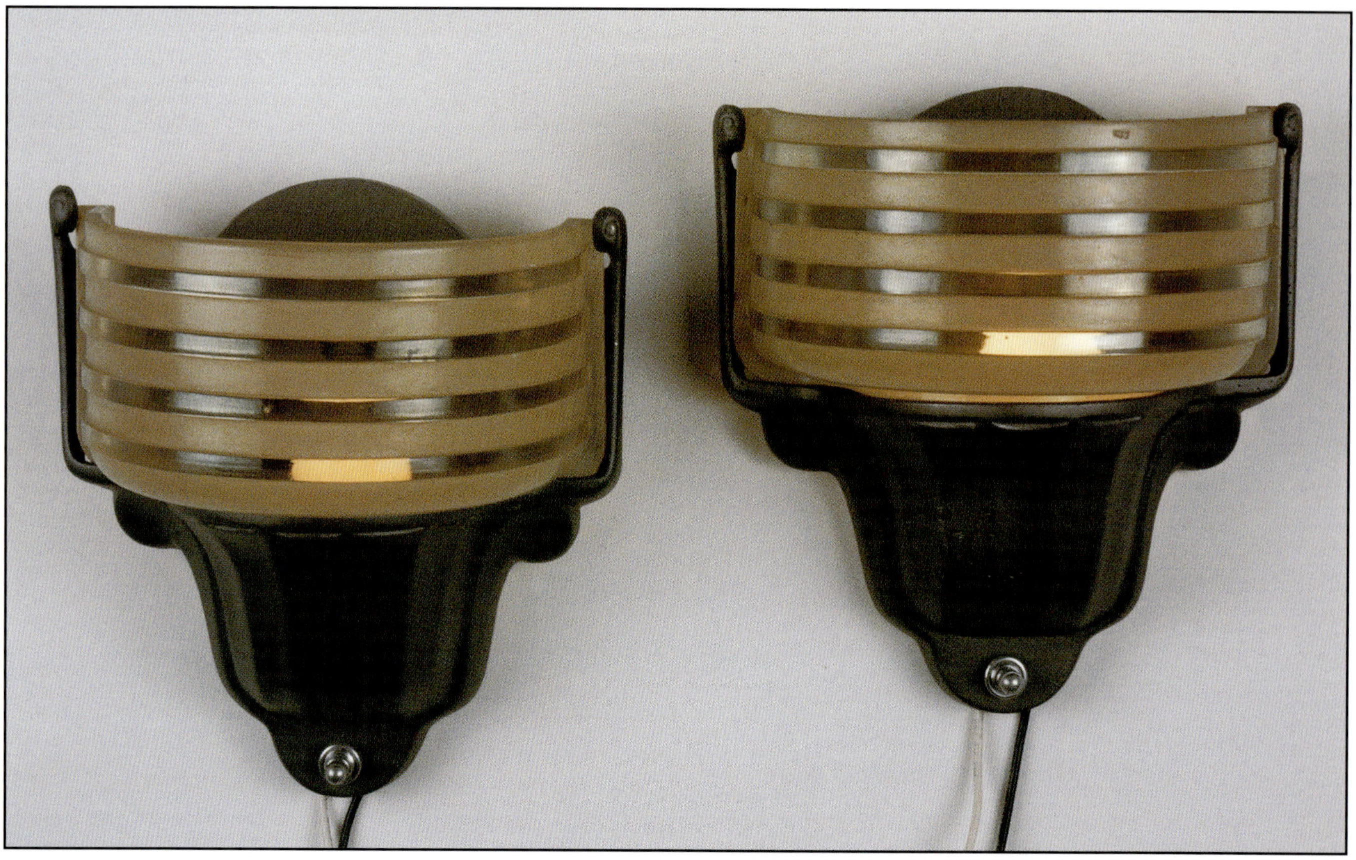

1930s art deco slip-shade wall sconces. Paper label reads "CSA Standard Electric." Composition: back plates are white metal with original gold-tone and pale green wash; slip-shades are white frosted with pink highlights and are shell-shaped. 8.5" H x 6" W x 4" protrusion from wall. $999-1100 pair.

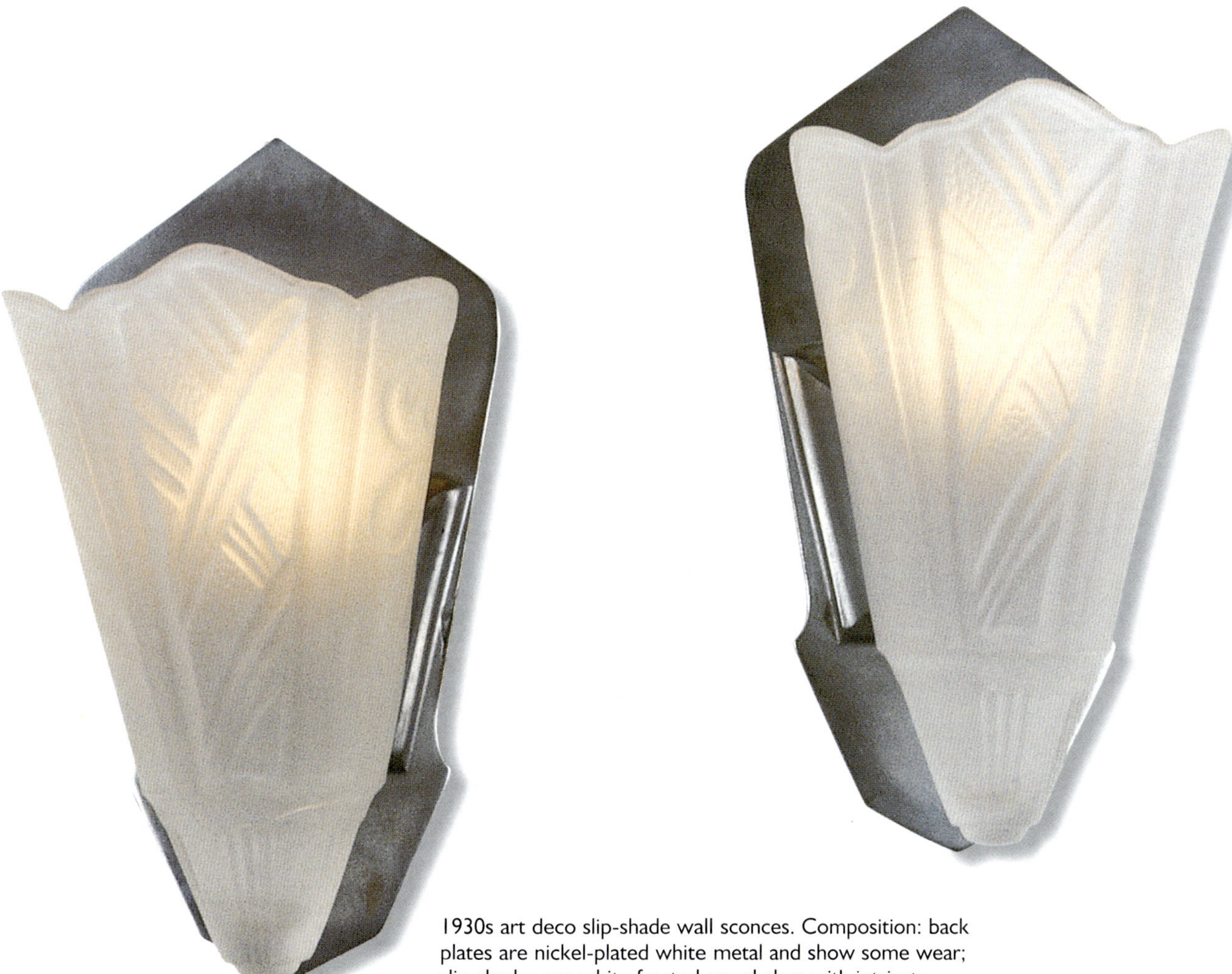

1930s art deco slip-shade wall sconces. Composition: back plates are nickel-plated white metal and show some wear; slip-shades are white frosted cased glass with intricate detail throughout. 9.75" H x 5" W x 6.25" protrusion from wall. $995-1100 pair.

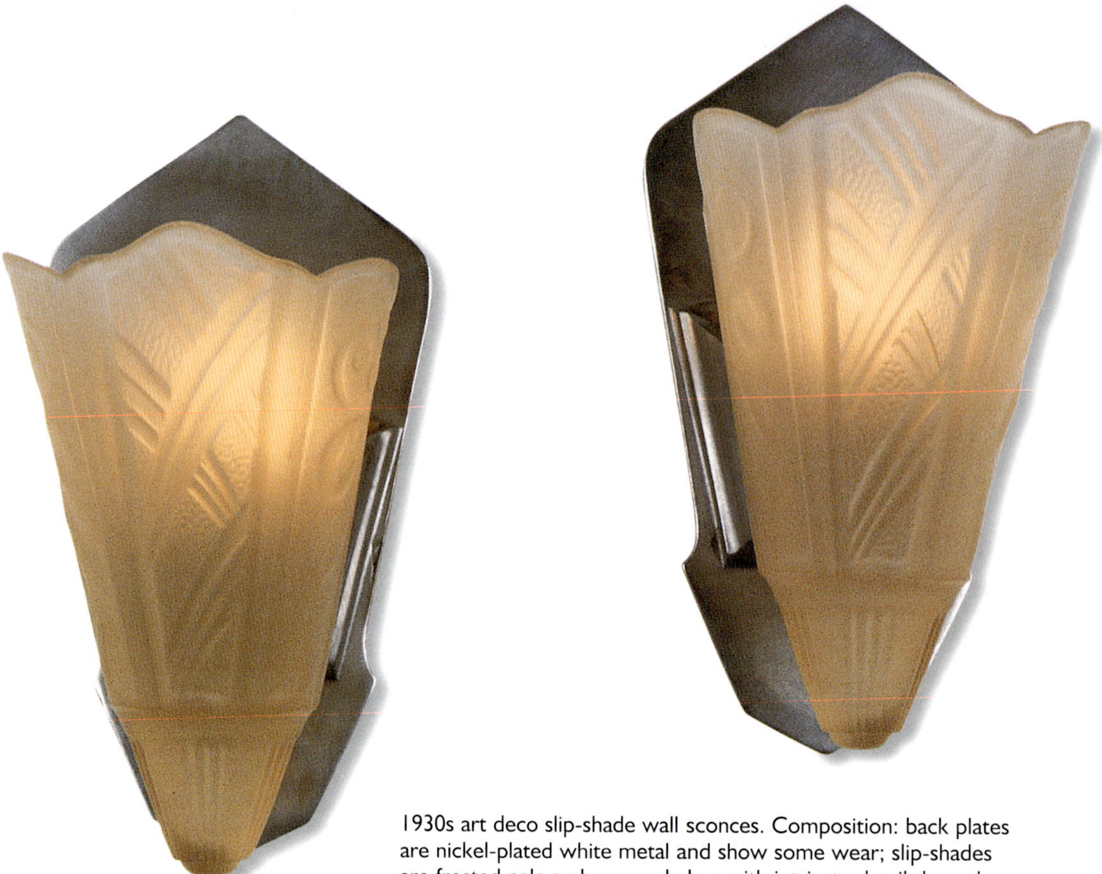

1930s art deco slip-shade wall sconces. Composition: back plates are nickel-plated white metal and show some wear; slip-shades are frosted pale amber cased glass with intricate detail throughout. 9.75" H x 5" W x 6.25" protrusion from wall. $995-1100 pair.

1930s art deco slip-shades wall sconces "Virden Manufacturing Co." Composition: white metal with gold-tone wash; slip-shades are amber pressed glass with floral design at center. 10.25" H x 6.25" W x 3.5" protrusion from wall. $965-1050 pair.

1930s art deco slip-shade wall sconces "Virden Manufacturing Co." Composition: white metal with bronze wash; slip-shades are pale amber pressed glass with floral design at center. 10.25" H x 6.25" W x 3.5" protrusion from wall. $965-1050 pair.

1920s art deco wall sconces manufactured by Bireley. Sconces used as promotion for Bireley orange drink. Composition: back plates are white metal with pewter wash, black highlights, and an intricate Deco step design. Bottles are pressed glass with original painted interior. 12.5" H x 4.75" W x 5" protrusion from wall. $1350-1895 pair.

1930s art deco slip-shade wall sconces. Composition: white metal with pewter wash and black highlights; slip-shades are cylindrical pressed glass, two-toned, light amber fading to brown. 11" H x 4.5" W x 5.5" protrusion from wall. $835-995 pair.

1930s art deco slip-shade wall sconces. Composition: white metal with gold-tone wash, white satin frosted shades. Back plates are signed "N.E.R. Inc." 12" H x 4.5" W x 6.25" protrusion from wall. $895-975 pair.

1930s art deco slip-shade wall sconces. Composition: back plates are white metal with bronze wash; slip-shades are a pale green frosted case glass. 9.75" H x 6.25" W x 4.5" protrusion from wall. $965-1050 pair.

1930s art deco wall sconces. Composition: back plates are chrome; shades are white milk glass with chrome accents at the top. 7.75" H x 3.75" W x 5" protrusion from wall. $495-585 pair.

1930s art deco slip-shade wall sconces, signed "Lightolier." Original paper label on back plate. Composition: white metal with bronze wash, original pull chain; slip-shades are curved cased glass with hand-painted grapes at center and amber border. 9.75" H x 5.75" W x 3.5" protrusion from wall. $925-1050 pair.

1930s art deco slip-shade wall sconces. Composition: white metal with cream colored paint and gold highlights; cased glass slip-shades are two-tone, amber and caramel in color. 11" H x 5.5" W x 4.5" protrusion from wall. $995-1085 pair.

1930s art deco slip-shade wall sconces with paper label, "Lightolier." Composition: back plates are chrome with milk glass shades. 4.25" H x 4.25" W x 8.25" protrusion from wall. $685-725 pair.

1930s art deco slip-shade wall sconces. Composition: back plates are white metal with pewter wash; slip shades light frosted cased glass with vertical design throughout. 9" H x 5.5" W x 6.25" protrusion from wall. $985-1095 pair.

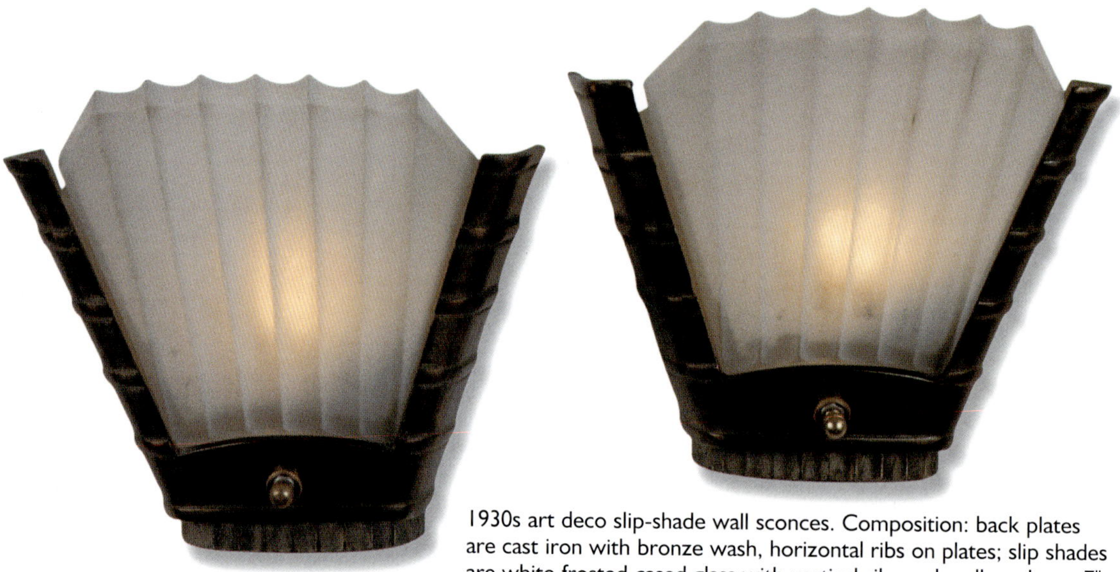

1930s art deco slip-shade wall sconces. Composition: back plates are cast iron with bronze wash, horizontal ribs on plates; slip shades are white frosted cased glass with vertical ribs and scalloped top. 7" H x 7.25" W x 4.25" protrusion from wall. $999-1085 pair.

1930s art deco slip-shade wall sconces. Composition: solid bronze with vertical ornamentation at base and ring; slip-shades are white frosted satin glass. 9" H x 6.25" W x 4" protrusion from wall. $1065-1200 pair.

1930s art deco slip-shade wall sconce. Composition: back plate is white metal with pewter wash; slip-shade is amber satin glass, circular with floral design throughout. 11.5" H x 4.5" W x 6.75" protrusion from wall. $465-525.

1930s art deco slip-shade wall sconces. Composition: back plates are white metal with chrome and pewter wash and copper accents; slip-shades are multicolored frosted satin glass, ranging from white frost to amber, vertical design. 11" H x 4.5" W x 4.5" protrusion from wall. $895-1050.

1930s art deco slip-shade wall sconces. Composition: back plates are white metal with chrome and pewter wash and copper accents; slip-shades are white frosted satin glass with art deco design throughout. 11" H x 4.5" W x 4.5" protrusion from wall. $895-1050 pair.

Early 1930s art deco slip-shade wall sconces. Composition: back plates are white metal with pewter wash signed "GLE Manufacturing Co." Slip-shades are very unusual amber and brown frosted glass; side panels mirror Chrysler Building. 8.5" H x 4.5" W x 4" protrusion from wall. $975-1150 pair.

1930s art deco wall sconces. Composition: back plates are white metal with pewter and copper wash; shades are tricolor amber, dark brown, and off-white. 6.5" H x 6.5" W x 8" protrusion from wall. $465-510 pair.

1930s art deco wall sconces. Composition: solid brass back plates with subtle ornamentation; shades are two-tone pressed glass, clear and off-white. 7" H x 5" W x 6.25" protrusion from wall. $395-445 pair.

1930s art deco wall sconces. Composition: back plates are nickel-plated white metal; shades are white frosted glass and have been replaced. Inner rings holding shades have also been replaced. 10.25" H x 7" W x 9.75" protrusion from wall. $765-825.

1920s art deco wall sconce. Composition: brass with slag glass panels that have been replaced. Appears to have come from a commercial venue. 16.5" H x 9.75" W x 6.25" protrusion from wall. $565-695.

1930s art deco ceiling fixture. Composition: white metal with white plastic band. Center ribbed glass lens at lower portion of fixture. 22" H x 24" D (shade). $965-1125.

1930s art deco slip-shade ceiling fixture. Composition: nickel-plated white metal; slip-shades are white frosted satin glass. 14" H x 8.74" W. $595-685.

Early 1930s art deco slip-shade ceiling fixture, manufactured by Moe Bridges. Composition: cast iron with bronze wash. 20" H x 9.75" D. $995-1145.

1930s art deco slip-shade ceiling fixture. Composition: frame is white metal with bronze wash; slip-shades are cylindrical amber cased satin glass with intricate detail. 15" H x 14" W. $575-695.

Early 1930s art deco slip-shade chandelier. Composition: the frame is white metal and brass; very unusual with intricate art deco detail throughout. Circular off-white cased glass slip-shades have unusual detail. All six of the tassels of the bottom of the chandelier are period. 18" H x 18.25" D. $1695-1850.

Early 1930s art deco slip-shade chandelier. Composition: frame is white metal and is very architectural in design; bands of Bakelite separate each slip-shade. Slip-shades were manufactured by "Consolidated Glass Company" and are pale amber and clear in color. A very hard to find chandelier. 27" H x 16.5" D. $2495-2800.

1930s art deco chandelier. Composition: frame is hammered aluminum; opalescent glass slip-shades fade from off-white to pale amber. Aluminum bands are at the top of each shade. 17" H x 18.5" D. $1695-1785.

1920s gothic chandelier. Composition: wrought iron and cast iron. Chandelier houses seven panels of original frosted glass. 14.5" H x 15.5" D. $695-850.

1930s art deco aluminum ceiling fixture with mirrored reflective interior to bounce light off of the ceiling. 33" H x 18.75" D. $645-765.

1930s art deco slip-shade ceiling fixture. Composition: white metal with bronze wash, intricate detail throughout frame, very architectural in design. Slip-shades are white frosted cased glass. 16.5" H x 16" span. $495-585.

1930s art deco ceiling fixture. Flush mount fitter has been replaced. Original pressed glass shade in clear and pink. 6.5" H x 12.5" D. $295-325.

1930s art deco pendant ceiling fixture. Composition: fitter is white metal with bronze wash. Pendant is white milk glass with art deco black enamel design throughout. 23" H x 11.5" D. $495-585.

1930s art deco pendant ceiling fixture. Composition: fitter is white metal with bronze wash. Pendant is white milk glass with black enamel art deco design throughout. 18.5" H x 11.5" D. $495-585.

1930s art deco milk glass school house fixture. Composition: fitter is white metal with bronze wash. 19" H x 12.5" D. $350-425.

1930s art deco milk glass ceiling fixture. Unusual design with the upper portion fading to clear. Composition: fitter is white metal with bronze wash. 15" H x 12.5" D. $295-325.

1930s art deco pendant ceiling fixture. Composition: fitter is solid brass. Pendant is white milk glass with black enamel stepped design. 14.25" H x 11.75" D. $365-410.

1920s art deco pendant ceiling fixture. Composition: fitter is white metal with a pewter wash; shade is amber crackle pyramid glass. A very hard to find fixture. 13" H x 9.5" D. $465-510.

1930s art deco milk glass pendant ceiling fixture. Composition: fitter is white metal with nickel plating. Pendant is white milk glass with black enamel stepped design. 4" H x 5.75" D. $295-345.

1930s art deco ceiling fixture. Composition: fitter is white metal with chrome plating. Shade is light blue opalescent cased glass with rings at base. 12.25" H x 10.5" D. $295-375.

1930s art deco ceiling fixture. Composition: copper and brass fitter with amber slag glass side panels and turquoise slag glass at base. 12.1" H x 5.5" D. $295-365

Early 1930s art deco slip-shade ceiling fixture manufactured by Lincoln. Composition: frame is white metal and hammered aluminum. Original pull chain at base. Slip-shades are amber cased glass with intricate design throughout. 12" H x 15.5" splay. $495-595.

1930s art deco ceiling fixture. Composition: fitter is solid brass. Globe is carnival glass with floral overlay and brass finial at base. 20.5" H x 14" D. $695-755.

Early 1920s art deco five-light chandelier, gothic in design. Composition: cast iron and bronze with original paint finish. 18" H x 20.5" D. $765-845.

1930s art deco five-light chandelier. Composition: wrought and cast iron with pewter wash. Chandelier has period canopy. 20.5" H x 18" D. $645-710.

Early 1920s art deco slip-shade chandelier. Composition: cast iron and white metal with industrial design throughout. Frame has pewter wash with brass highlights. Chandelier houses six lights. The five slip-shades and bottom plate manufactured by Consolidated Glass Company. Shades are pale amber with clear floral designs throughout. 19" H x 17.25" D. $2350-2900.

1930s art deco three-light slip-shade ceiling fixture. Composition: frame is white metal with glass florets. Frame has original cream enamel finish. Slip-shades are pressed glass in a combination of clear and cream color. 7.5" H x 14.5" D. $365-450.

Early1930s art deco slip-shade flush mount chandelier. Composition: white metal and bronze with intricate detail throughout frame. Slip-shades are very unusual in design and are amber fading to brown at the corners, reminiscent of Radio City and the Chrysler Building. 10.5" H x 18" D. $985-1125.

Early 1930s art deco seven-light slip-shade chandelier. Composition: bronze with architectural center shaft and frame. Five slip-shades and bottom plate are white cased glass with clear overlay. 21" H x 17" D. $1699-1850.

Early 1930s art deco seven-light, slip-shade chandelier, architectural design throughout the frame and canopy. Composition: nickel-plated white metal. Five slip-shades and bottom plate manufactured by Consolidated. Shades are pale yellow with clear floral design. 19" H x 18.5" D. $2250-2800.

Early 1940s art deco chandelier. Composition: chrome-plated frame with five circular pale amber shades. 16" H x 18" D. $495-585.

1930s art deco slip-shade chandelier. Composition: frame constructed of white metal with bronze wash; original tassels at base of chandelier. Oval opalescent glass shades with art deco design embossed on front of each shade. 16" H x 19" D. $1695-1975.

Late 1920s art deco slip-shade chandelier. Composition: white metal and cast iron with intricate art deco detail throughout exterior and interior of frame. Chandelier frame is signed "Williamson-Chicago" on each panel that houses a slip-shade. Slip-shades are white cased glass with pale yellow vertical strip. 17.5" H x 15" D. $2395-2800.

1930s art deco slip-shade chandelier. The chandelier has five unusual slip-shades and bottom plate. Composition: the frame is cast iron and white metal and has its original cream and amber baked enamel finish. All of the shades are cased glass in pale amber with vertical ribs and two horizontal stripes in brown. 14" H x 15.5" D. $1399-1650.

1930s art deco three-light slip-shade fixture. The frame is very architectural in design. Composition: white metal with pewter and brass wash; slip-shades are white frosted cased glass with scalloped edges at top. 12" H x 12.5" D. $849-985.

1930s art deco ceiling fixture. This fixture obviously came from a church. Composition: brass, original panels of clear glass. The fixture houses five lights with a hexagonal grid at the base. 28" H x 11" D. $1125-1300.

1930s art deco six-light slip-shade chandelier. Composition: white metal with bronze wash; ornamentation of Saturns and planets on upper and lower portions of the center shaft. The six slip-shades are off-white cased glass with a crest at the center of each shade. 22" H x 18" D. $1599-1700.

1920s art deco ceiling fixture. Shade is Phoenix glass, off white in color. Brass fitters and chain have been replaced. 9" H x 16.25" D. $985-1095.

1930s art deco slip-shade chandelier. Composition: frame is white metal with gold-tone wash and pressed glass accents. Shades are circular off-white cased glass, very unusual in design. Separate on/off switch at the base of the frame. 14" H x 18.5" D. $985-1050.

1930s art deco slip-shade chandelier. Composition: frame is white metal with gold-tone wash; shades are frosted white cased glass with vertical lines throughout. 18" H x 16.5" D. $1165-1295.

1930s art deco slip-shade chandelier. Starburst at center of the frame repeated in the ceiling canopy. Composition: white metal with pewter and bronze wash. Separate on/off switch at base. Slip-shades are amber in color with intricate art deco details. 15" H x 17.5" D. $999-1250

1930s art deco slip-shade chandelier. Composition: white metal with bronze and pewter wash with glass accents; slip-shades are concave, off-white cased glass with deco designs throughout each shade. 15" H x 15.5" D. $1165-1295.

1930s art deco six-light slip-shade chandelier. Composition: frame is white metal with pewter wash and very industrial in design. Five slip-shades and bottom plate are white frosted glass with slight pink tint. Shades are cased glass with scallops at upper portion. 16" H x 15.25" D. $1095-1250.

1930s art deco slip-shade chandelier. Composition: white metal with bronze and pewter wash. Slip-shades are arched white frosted cased glass with vertical design. Separate on/off switch at base of frame. 15" H x 16.5" D. $1165-1300.

1930s art deco slip-shade chandelier. Composition: frame is white metal with bronze wash. Slip-shades are white frosted cased glass with vertical design throughout. 17.5" H x 14" D. $1165-1325.

Early 1930s art deco ceiling fixture. Composition: frame is white metal with intricate art deco detail throughout. Portions of the frame are nickel-plated, and it has brass insets. Original glass panels at center. Fixture has five exposed bulbs facing downward. 23" H x 19" D. $965-1200.

1920s French art deco ceiling fixture. Composition: white metal with intricate art deco detail throughout the frame. Five jadite slip-shades are stepped and are period. 15" H x 20" D. $1695-1900.

1930s art deco triple halophane ceiling fixture. Composition: housing is nickel-plated white metal. Original tag affixed to frame. 27" H x 13" D. $695-840.

1930s art deco ceiling fixture made in Germany. Composition: fitter is brass, showing some wear. Pendant is cream and white cased glass. 22.5" H x 7.75" D. $395-485.

1930s art deco slip-shade chandelier. Composition: frame is white metal with bronze wash, intricate detail throughout. Slip-shades are amber cased glass with floral design. 16" H x 15.5" D. $995-1100.

1930s art deco slip-shade chandelier. Composition: white metal with original painted finish, cream with tones of green and amber; silver leaf at center shaft. Frame boasts two arrows surrounded by a wreath. Slip-shades are amber cased glass, each having a wheat pattern at its center. 21" H x 15.5" D. $1295-1450.

1930s art deco three-light slip-shade chandelier. Composition: white metal with pewter wash and black highlights. Slip-shades are cylindrical amber cased glass with dark brown highlights. 12.5" H x 15.5" D. $785-900.

1930s art deco three-light ceiling fixture. Composition: white metal and brass frame; white metal has a baked cream finish. Shades are white milk glass with vertical design. 6" H x 12" D. $495-610.

1920s art deco pendant ceiling fixture made in Germany. Composition: fitter is white metal with pewter wash and original cloth wrap. Pendant is milk glass with chevron design in gray, orange, and peach colors. 24" H x 9" D. $395-515.

Late 1920s art deco ceiling fixture. Composition: fitter and canopy are bronze with intricate art deco detail. Pendant is pressed glass in ice blue iridescent color, four-tiered. 24" H x 13.5" D. $695-850.

1930s art deco slip-shade ceiling fixture. Frame manufactured by Lightolier. Composition: white metal with gold tone wash. Pink and clear glass slip-shades by Consolidated Glass. A rare find. 15" H x 6.75" W. $865-925.

1930s art deco slip-shade ceiling fixture. Frame manufactured by Lightolier. Composition: white metal with gold tone wash. Amber and clear glass slip-shades by Consolidated Glass. A rare find. 15" H x 6.75" W. $865-925.

1930s art deco slip-shade ceiling fixture. Frame manufactured by Lightolier. Composition: white metal with gold tone wash. Light green and clear glass slip-shades by Consolidated Glass. A rare find. 15" H x 6.75" W. $865-925.

1930s art deco slip-shade chandelier. Composition: frame is white metal with pewter wash and black and gold accents. Circular shades are frosted with caramel decorations. Also has separate on/off switch. 19" H x 17" D. $975-1125.

Late 1920s art deco slip-shade chandelier. Composition: cast iron with intricate art deco design. Six panels of ice glass have been replaced. 19" H x 15" D. $945-995.

1930s art deco seven-light slip-shade chandelier. Composition: frame is white metal with bronze wash. Intricate art deco detail throughout frame and canopy. Shades are milk glass with caramel colored accents. 19" H x 16" D. $1395-1500.

1930s art deco slip-shade chandelier, very architectural in design. Composition: white metal with pewter wash. Frame has separate on/off switch. Shades are milk glass with pale amber design throughout. Shades are also quite architectural in design .16" H x 17.5" D. $1199-1450.

1930s art deco slip-shade chandelier. Composition: frame is white metal with bronze wash and original paint. This is a flush mount chandelier. Slip-shades are pale amber satin glass. 14.5" H x 15.75" D. $895-1125.

Early 1930s art deco six-light slip-shade chandelier. Frame manufactured by Lightolier. Very architectural in design. Composition: White metal with bronze wash. Five slip-shades and bottom plate manufactured by Consolidated Manufacturing Co. Shades are amber in color with clear floral designs throughout. 19" H x 14" D. $1499-1750.

1930s art deco seven-light slip-shade chandelier. Composition: frame is white metal with architectural shaft design, gold tone wash. Slip-shades and bottom plate are white satin glass. 19" H x 16" D. $1395-1550.

1930s art deco ceiling fixture. Composition: white metal and aluminum with cream-colored baked enamel finish. Fixture also has chrome accents. 20" H x 16" D. $595-675.

Late 1920s art deco seven-light slip-shade chandelier. Composition: frame is very architectural in design, constructed of white metal with a pewter wash. Five slip-shades and bottom plate are milk glass with silver horizontal bands throughout. Frame has separate on/off switch at base. 19" H x 17" D. $1999-2300.

Early 1930s art deco ceiling fixture. Composition: shaft is chrome-plated with sphere at center; pendent is circular milk glass with baked art deco design throughout. 32" H x 12.5" D. $595-725.

Early 1930s art deco ceiling fixture. Composition: fitter and ceiling canopy are solid brass with original chain. Pendant is pressed glass with intricate design throughout, pale blue in color. 30.5" H x 11.5" D. $595-700.

1930s art deco slip-shade chandelier. Composition: frame is white metal with gold tone wash; slip-shades are pale yellow satin cased glass. 19" H x 18" D. $965-1125.

1930s art deco slip-shade ceiling fixture. Composition: frame is ceramic in pale yellow with gold accents. The two slip-shades are dark amber cased glass. 13" W x 7.5" Depth x 4.5" protrusion. $499-599.

1930s art deco slip-shade ceiling fixture. Composition: frame is ceramic in pale yellow with gold accents. The three slip-shades are dark amber cased glass. 12.5" D x 4.5" protrusion. $525-710.

1930s art deco slip-shade chandelier. Composition: frame is white metal with pewter wash and black accents, flush mount frame. Slip-shades are frosted satin cased glass. 14" H x 12.5" D. $975-1200.

Early 1930s art deco seven-light slip-shade chandelier. Composition: white metal with copper wash. Center shaft boasts starburst. Five slip-shades and bottom plate are pale amber satin cased glass. 22" H x 15.5" D. $1499-1750.

1930s art deco two-light slip-shade ceiling fixture. Composition: fitter is white metal with baked cream color finish and gold accents. Slip-shades are pink cased glass. 17" W x 6" D x 5" protrusion. $499-595.

1930s art deco slip-shade chandelier. Composition: frame is bronze with intricate detail throughout. Five slip-shades are custard color with caramel accents throughout each shade. 13" H x 14.5" D. $1799-1950.

1930s art deco pendant ceiling fixture. Composition: white milk glass with baked enamel design throughout; lower portion is pressed glass with three tiers and a chrome-plated band holding it in place. Fitter is white metal with bronze wash. 15.5" H x 12.25" D (pendant). $695-800.

1930s art deco light fixture. Composition: original amber slag glass with green slag glass at base. Frame is white metal with copper finish and brass highlights. 13" H x 5.75" D. $495-600.

1930s art deco ceiling fixture came from synagogue in Boston, Massachusetts. Composition: frame is white metal with gold tone wash, decorated with two Stars of David and two Menorahs. White milk glass interior. 24" H x 17" D. $975-1150.

1930s art deco chandelier. Composition: frame is chrome-plated white metal. Shades are opalescent glass. 18" H x 22" D. $695-785.

Late 1920s art deco seven-light slip-shade chandelier. Composition: frame is bronze with intricate art deco detail throughout. Five slip-shades and bottom plate are two-tone amber cased glass. 24" H x 20" D. $1995-2350.

1930s art deco slip-shade chandelier. Composition: frame is white metal with bronze wash; pale amber cased glass shades. 19" H x 17.75" D. $1150-1400.

1930s art deco three-light slip-shade ceiling fixture. Composition: frame is white metal with three circular custard cased glass shades. 14.5" H x 15.5" D. $785-900.

Early 1930s art deco copper ceiling fixture from a Catholic church. Fixture houses a circular grid at the base. 18" H x 12.25" D. $695-750.

1930s art deco slip-shade chandelier. Composition: frame is white metal with a bronze wash. Slip-shades are amber satin cased glass. 17.5" H x 18" D. $1199-1450.

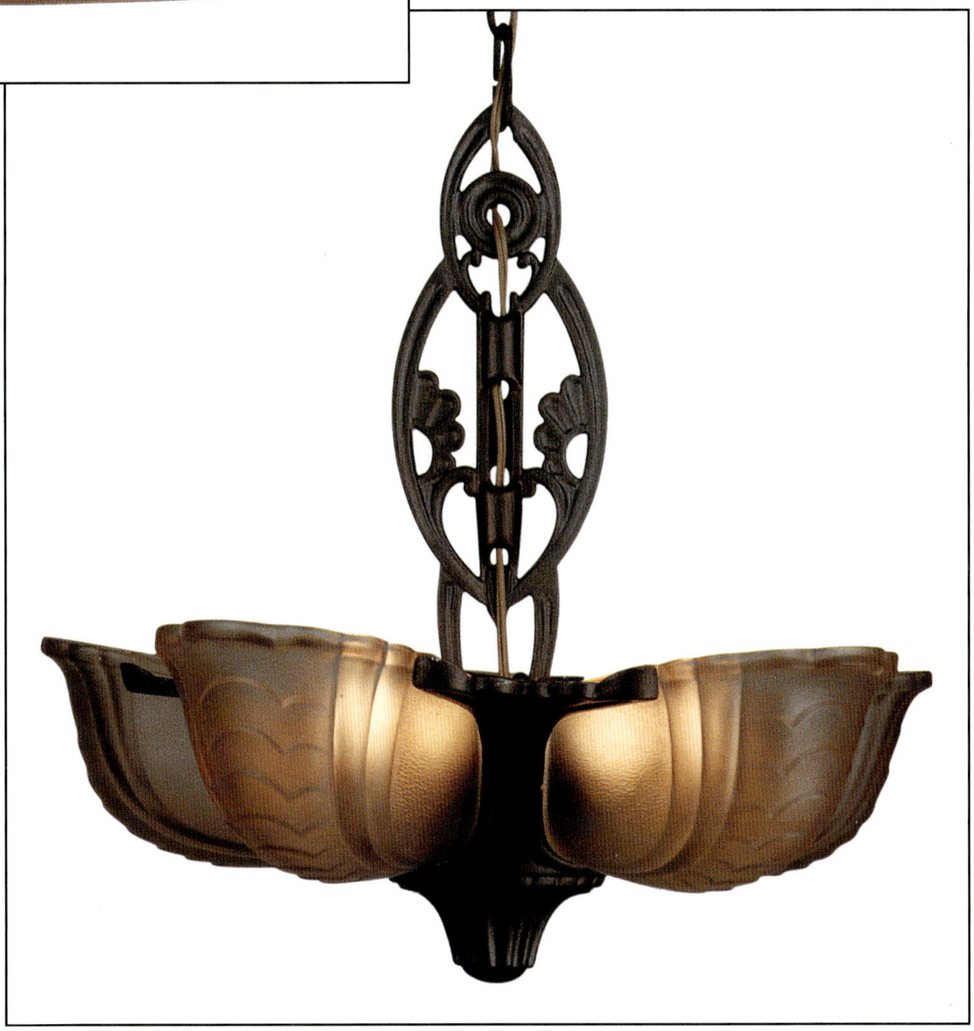

1930s art deco slip-shade chandelier manufactured by Lincoln Manufacturing Co. Composition: frame is white metal with pewter wash and gold tone highlights. Separate on/off switch above dangle at base. Five slip-shades are amber cased glass with floral design on each. 21" H x 17.75" D. $1199-1299.

Early 1930s art deco slip-shade chandelier manufactured by "Beardslee-Chicago." Composition: frame is white metal with pewter wash. Five circular slip-shades in pale amber cased glass with floral design throughout each shade. 20" H x 15.5" D. $1399-1550.

1930s art deco slip-shade chandelier. Composition: white metal with bronze wash. Slip-shades are white frosted cased glass with floral design on each shade. 17.5" H x 18" D. $1499-1650.

Early 1930s art deco slip-shade chandelier. Composition: frame is white metal with bronze wash and chrome highlights. Five panels of mica separate the five pale amber cased glass slip-shades. They have dark amber accents at the corners. Intricate details throughout shades and frame. 19" H x 15" D. $2295-2650.

1930s art deco church ceiling fixture. Composition: frame is solid brass with original frosted bubble glass and milk glass interior reflector. 20" H x 12" D. $695-785.

1930s art deco slip-shade chandelier. Composition: white metal and chrome with bronze wash. Slip-shades are heavy cased glass in a cream color with vertical ribs throughout each shade. 13" H x 13.75" D. $1199-1400.

1930s art deco slip-shade chandelier. Composition: white metal with bronze wash. Slip-shades are white frosted cased glass with intricate detail throughout. 15" H x 16.5" D. $1299-1550.

1930s art deco slip-shade chandelier. Composition: frame is white metal with pewter wash. Slip-shades are three-tone peach cased glass. A very rare find. 26" H x 15" D. $1899-2200.

1930s art deco slip-shade three-light ceiling fixture. Composition: frame is white metal with baked enamel cream finish. Slip-shades are pressed glass, which are two-tone cream and clear. 6" H x 13.75" D. $499-610.

Late 1930s French art deco chandelier. Composition: frame is chrome and walnut. Four surrounding shades and bottom plate are cream cased glass with vertical ribs throughout each shade. 38" H x 29.5" D. $1325-1575.

1930s art deco slip-shade chandelier. Composition: frame is white metal with copper wash. Five slip-shades and bottom plate are white frosted cased glass with vertical ribs throughout. 19" H x 14.5" D. $999-1250.

Early 1930s French art deco chandelier. Composition: frame is copper and walnut with mahogany accents on shaft. Four shades and bottom plate are pressed glass with satin frosted accents. 32" H x 35" D. $1795-2150.

1930s art deco pendant ceiling fixture. Composition: fitter is chrome plated white metal. Pendant is white milk glass with black enamel designs and vertical ribs throughout. 14" H x 12" D. $595-695.

1930s art deco slip-shade chandelier. Composition: frame is chrome and white metal with a bronze wash. Five circular slip-shades are custard cased glass with an intricate design. 15" H x 19.25" D. $999-1199.

Early 1930s art deco slip-shade chandelier. Composition: frame is white metal and cast iron with a pewter wash. Five slip-shades are white frosted glass. Both the frame and shades are very industrial in design. 20" H x 18" D. $1399-1550.

Early 1930s art deco slip-shade chandelier. Composition: frame is white metal with copper wash. Intricate art deco design throughout the frame. Slip-shades are inverted, which are very unusual. Shades are a pale yellow cased glass with art deco design throughout each shade. 18" H x 15.5" D. $1299-1400.

Early 1930s art deco slip-shade chandelier. Frame is manufactured by "Lightolier." Composition: frame is white metal and cast iron with goldtone wash. Five slip-shades and bottom plate were manufactured by "Consolidated Glass Company." Slip shades are amber with clear floral designs throughout each. 24" H x 18" D. $1499-1750.

1930s art deco chandelier. Composition: frame is copper with chrome accents. Slip-shades are pale amber opalescent glass. 12" H x 19.5" D. $699-825.

Early 1930s French art deco chandelier. Composition: frame is solid brass with intricate floral and grape design on all sides. Original white frosted glass panels throughout the frame. 30" H x 36" W x 36" D. $2850-3400.

1930s art deco fixture houses three lights with intricate art deco detail throughout. Original pull chain intact. $295-375.

Early 1930s art deco wall sconces. Composition: chrome plated back plates with silver & cobalt blue mirror insets. Clear replacement rods serve as the shades for lights. 54" High x 14" W. "A Very Rare Find" $4200-4800 pair.

Floor Lighting

We can all remember seeing floor lighting when we went to visit our grandparents. Of course, they seemed so tall as we were only three to four feet high. Those of us age fifty and older remember growing up with these lights in our own homes.

The torchiere was once also considered a staple for a commercial venue, and now the torchiere has become a very popular art deco floor lamp in the twenty-first century.

Torchieres range in height from 4-1/2 feet to 6-1/2 feet. These lamps had porcelain mogul sockets installed. Mogul bulbs were three-way bulbs that ran up to 350 watts and the porcelain sockets could withstand the heat and not cause a fire in the unit. With a three-way bulb, torchieres could light a portion of a room or the entire room. The torchiere bases and shafts were manufactured with a variety of metals ranging from white metal to bronze and brass. Some were finished in a bronze wash, gold tone wash, brass-plated while others were chrome-plated. A number of torchieres were manufactured with a cast iron base so that the lamp would be bottom heavy and could not be easily overturned.

More elaborate torchieres had bases that were adorned with marble, alabaster, bakelite or celluloid. A few high-end torchieres were designed to have a lighted base that would have a separate on/off switch.

The torchiere shades came in a wide variety of shapes and sizes. They were manufactured with floral designs, geometric designs, and some were smooth glass with no design at all. They were made of cased glass and pressed glass. They came in a wide variety of colors and textures. Most popular were the pale amber shades that were in the shape of a large bowl.

Bridge lamps were designed to create a direct light, mostly for reading. The term bridge lamp was coined as the actual light socket and shade extends off of a bridge from the main shaft of the lamp. Many of the bridge lamps had an extensive amount of intricate detail throughout the base, shaft, and bridge. The bridge lamps were constructed from many of the same compositions that were used to manufacture the torchieres. However, shades on the bridge lamps ranged from decorative hand-painted glass shades to halophane glass, pressed glass, cased glass, parchment paper, and even fabric shades.

1930s art deco bridge lamp. Composition: white metal with period shade. 55.5" H x 9.5" D (base). $475-550.

1930s art deco bridge lamp. Composition: Akro Agate, brass, and cast iron with a bronze wash. Shade has been replaced with a French glass shade. 58" H x 9" W x 9" D. $985-1100.

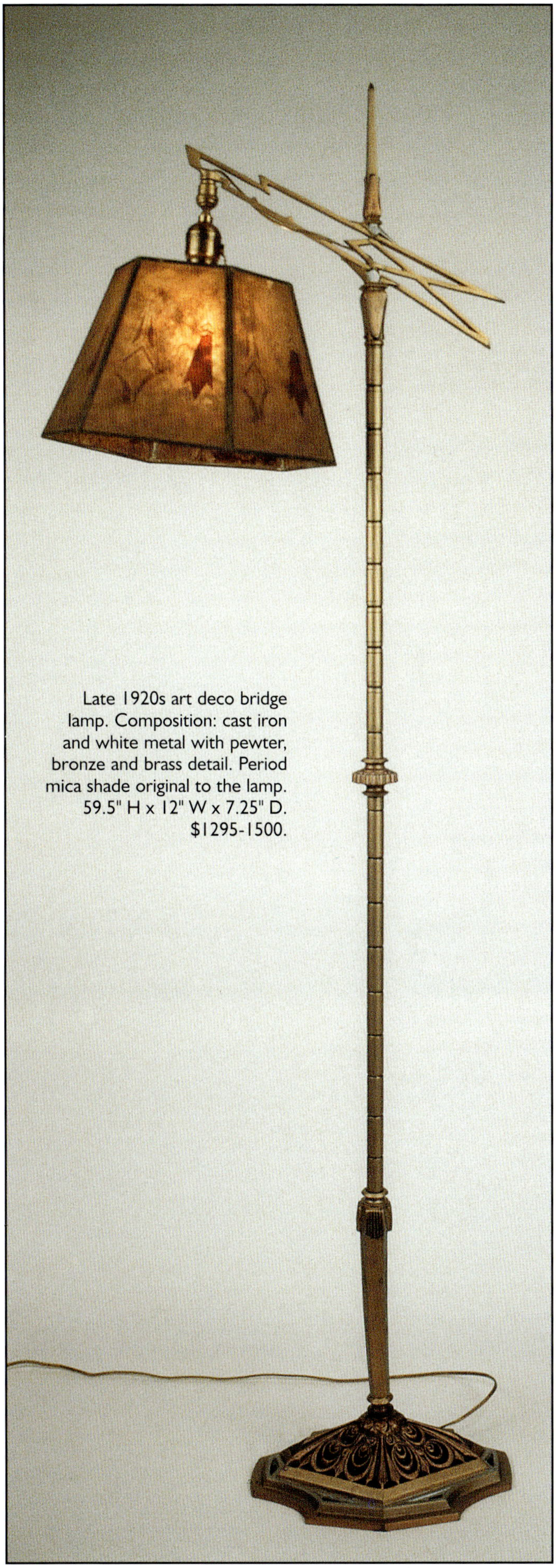

Late 1920s art deco bridge lamp. Composition: cast iron and white metal with pewter, bronze and brass detail. Period mica shade original to the lamp. 59.5" H x 12" W x 7.25" D. $1295-1500.

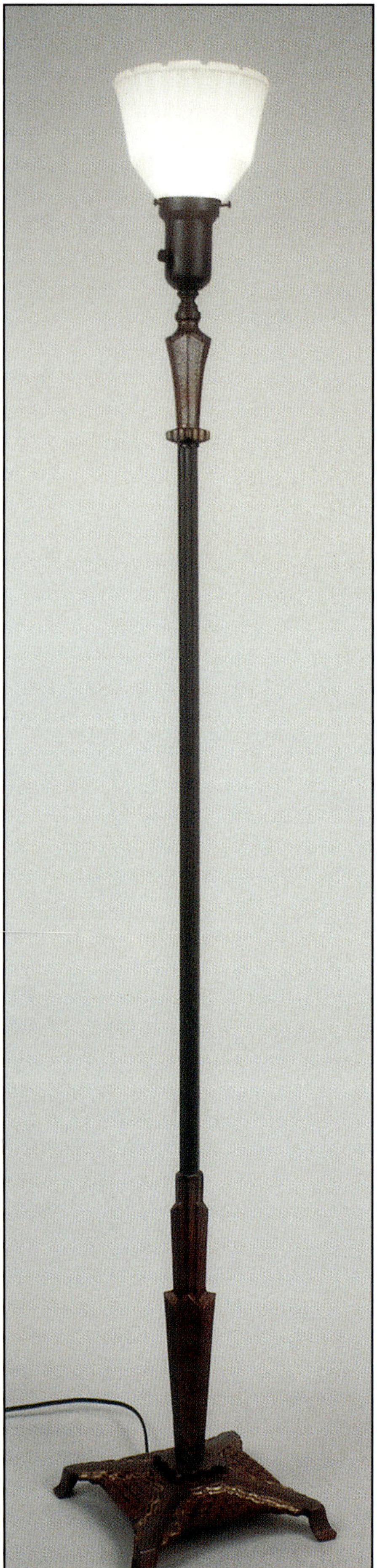

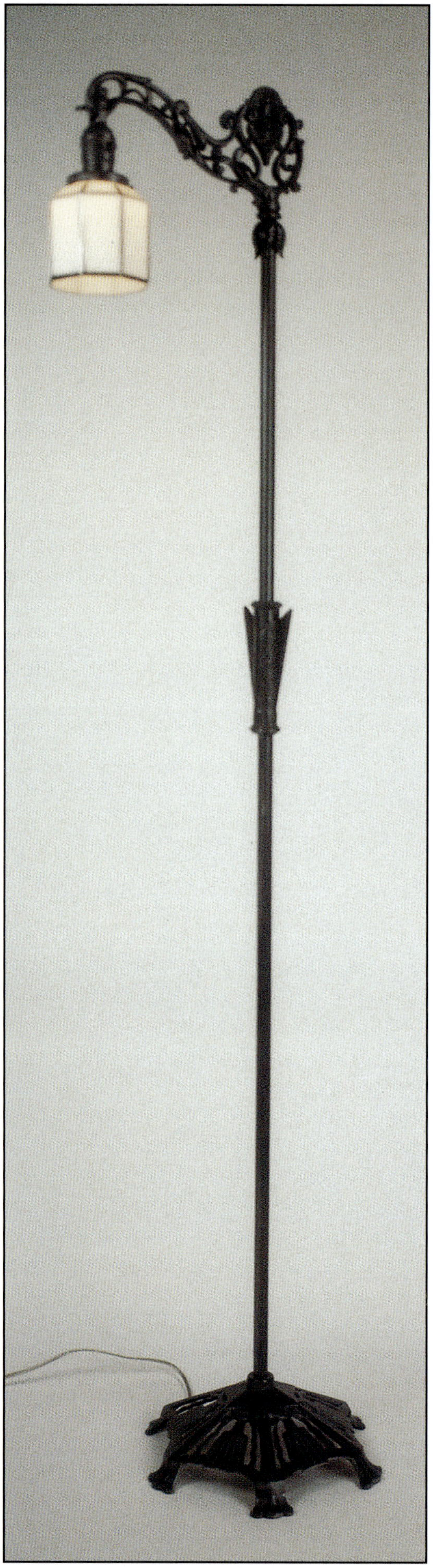

1930s art deco floor lamp. Composition: cast iron, white metal with red, gold, and pewter wash. Original pressed milk glass shade. 56.50" H x 8.75" W x 8.75" D. $465-550.

1930s art deco bridge lamp. Composition: white metal and cast iron. Great detail. Refinished with pewter wash, period shade. 58.5" H x 12" D (base). $465-550.

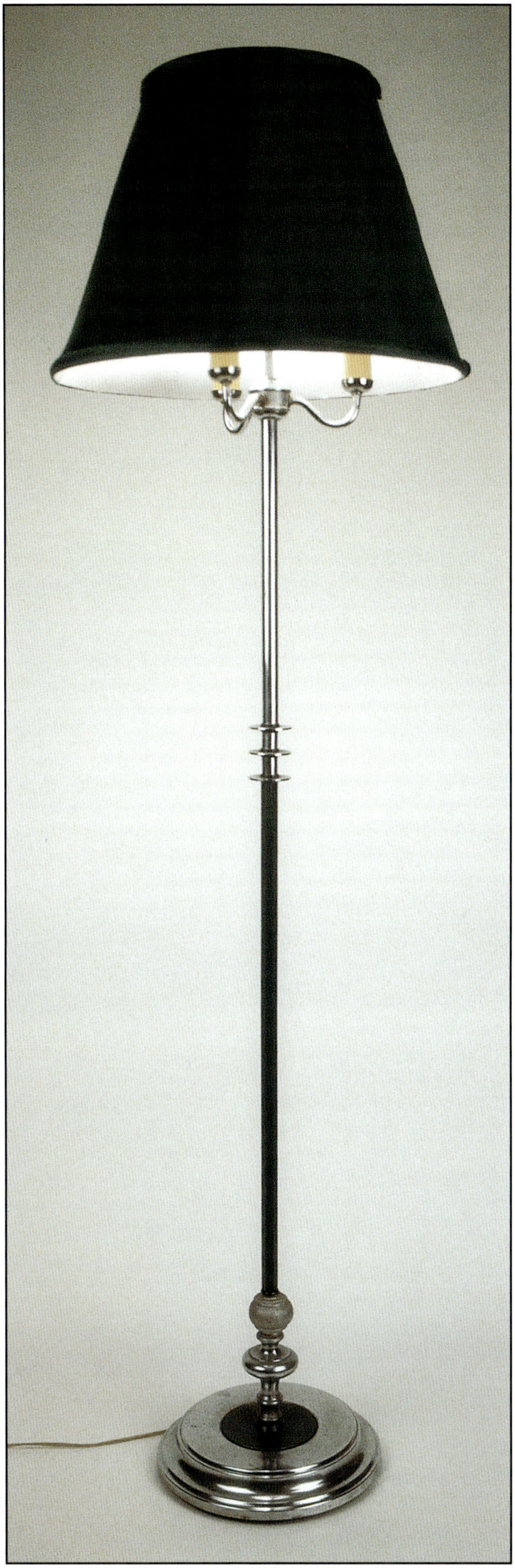

1930s art deco floor lamp. Composition: marble, cast iron, and white metal with bronze wash and green accents. Silk shade has been replaced. 66.5" H x 10" D (base). $585-700.

1930s art deco floor lamp. Composition: cast iron, chrome-plated with green accent. Silk shade and finial have been replaced. 64.5" H x 11.25" D (base). $610-700.

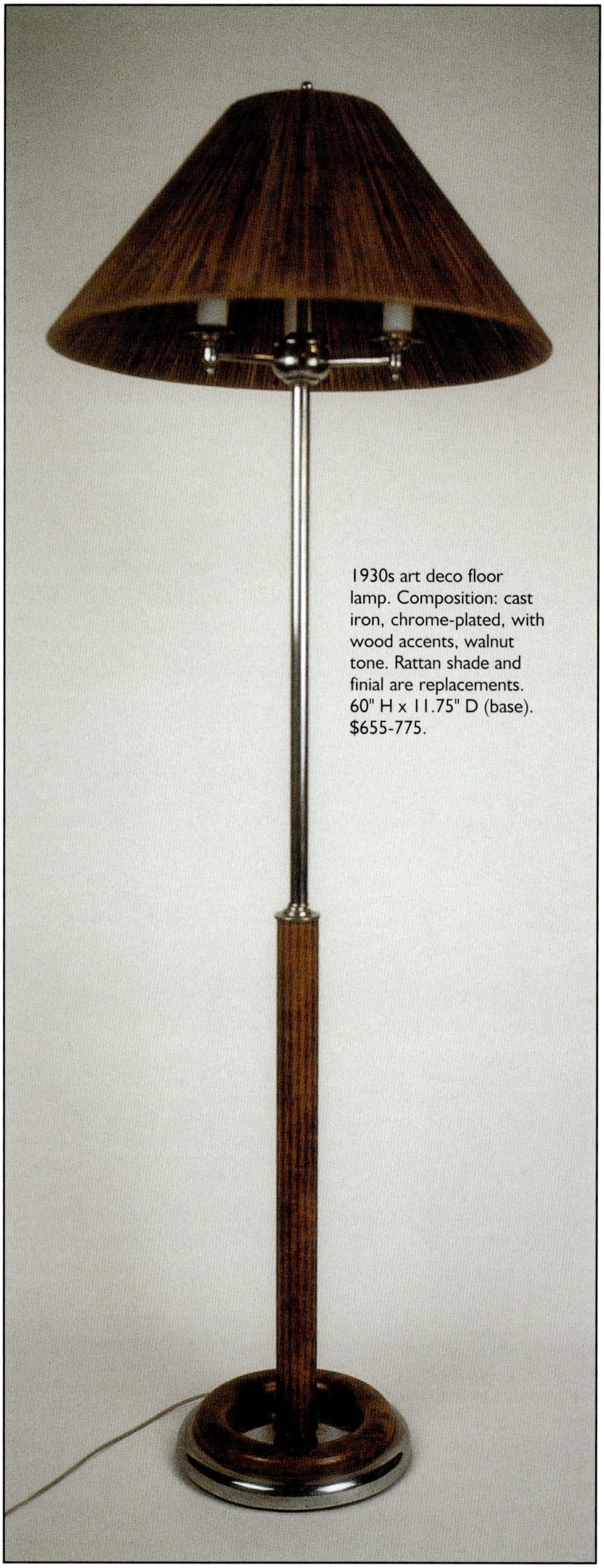

1930s art deco floor lamp. Composition: cast iron, chrome-plated, with wood accents, walnut tone. Rattan shade and finial are replacements. 60" H x 11.75" D (base). $655-775.

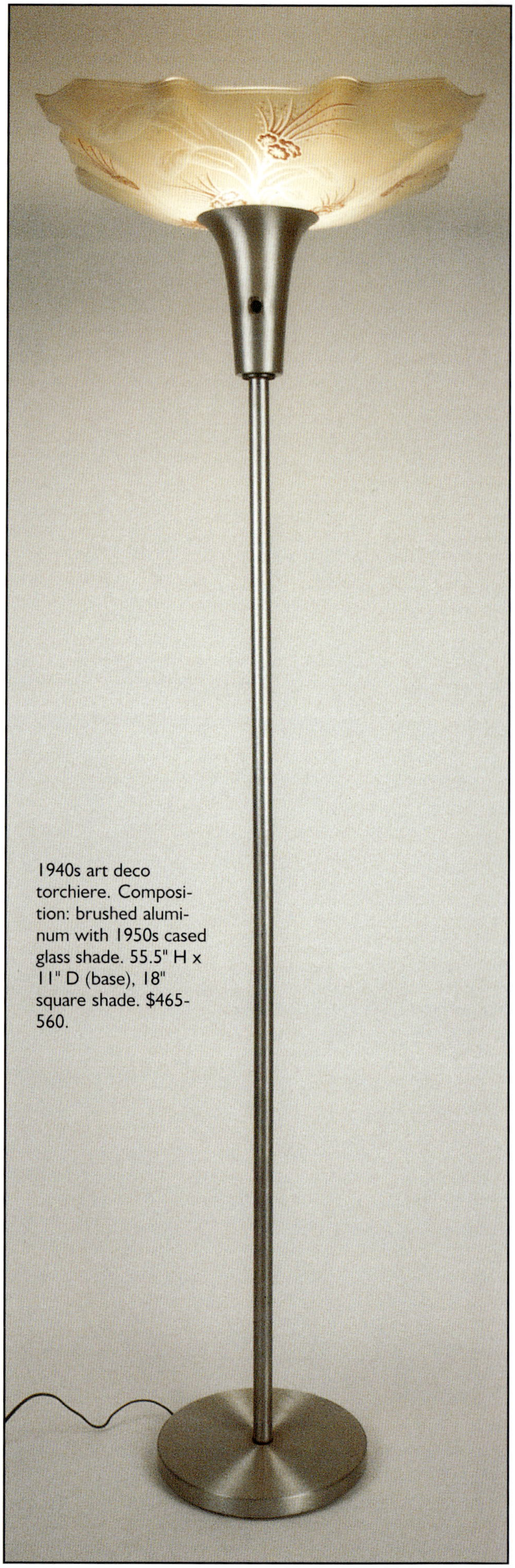

1940s art deco torchiere. Composition: brushed aluminum with 1950s cased glass shade. 55.5" H x 11" D (base), 18" square shade. $465-560.

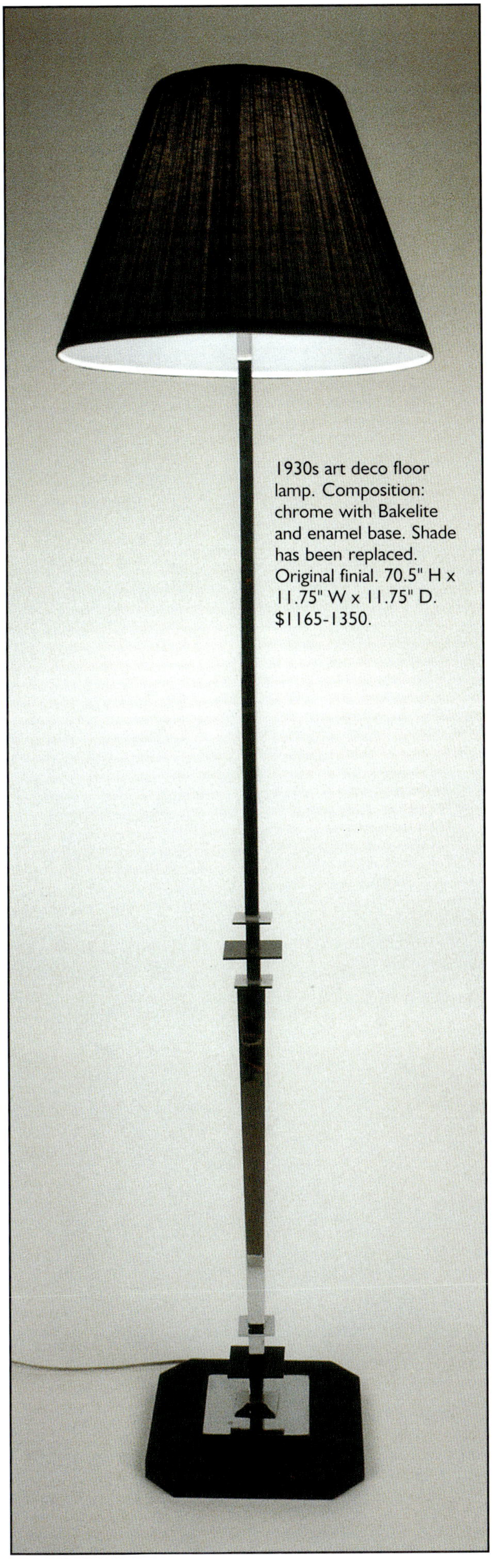

1930s art deco floor lamp. Composition: chrome with Bakelite and enamel base. Shade has been replaced. Original finial. 70.5" H x 11.75" W x 11.75" D. $1165-1350.

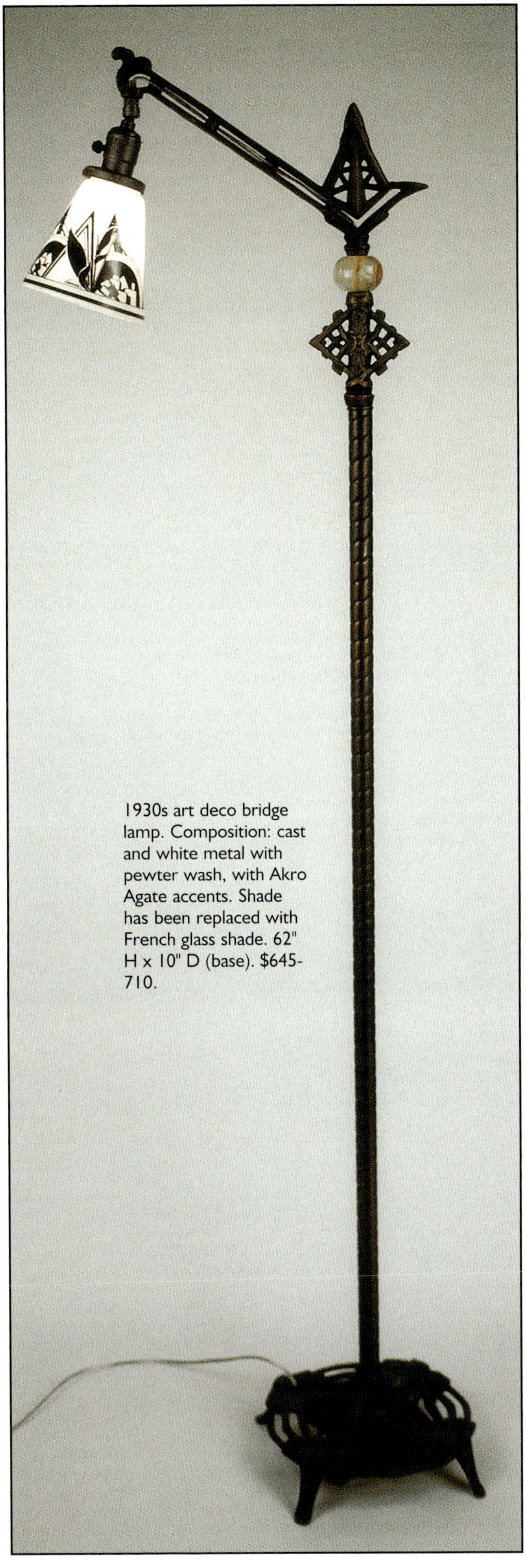

1930s art deco bridge lamp. Composition: cast and white metal with pewter wash, with Akro Agate accents. Shade has been replaced with French glass shade. 62" H x 10" D (base). $645-710.

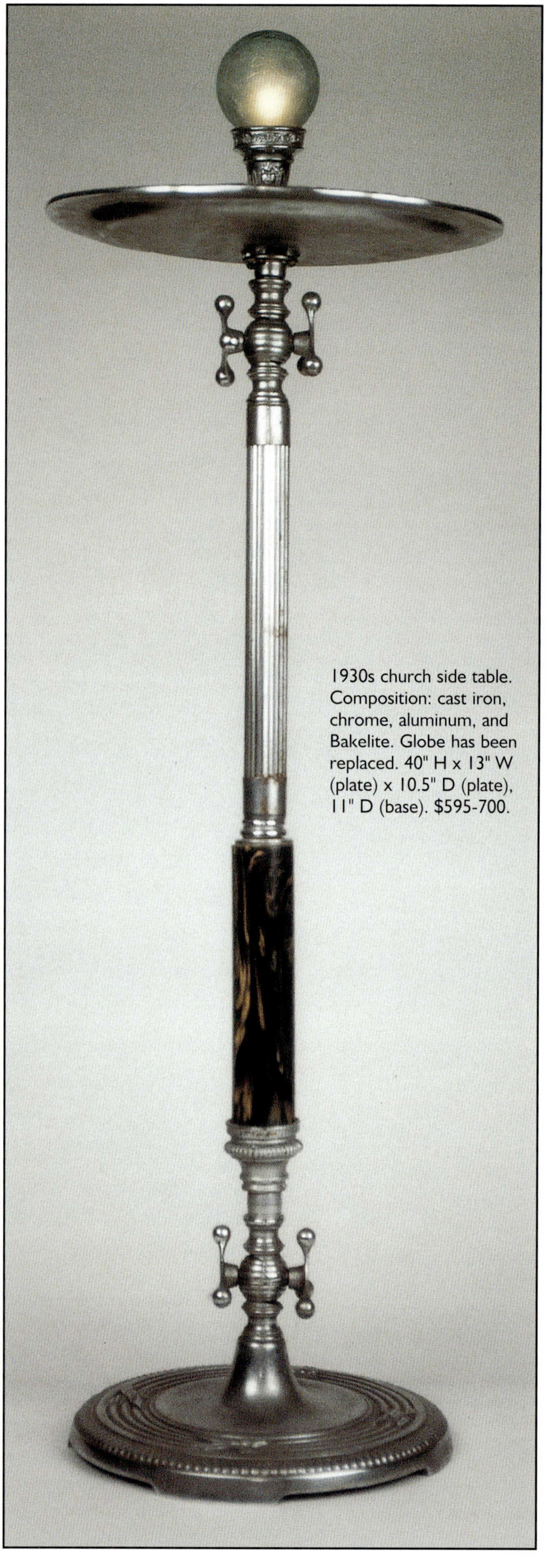

1930s church side table. Composition: cast iron, chrome, aluminum, and Bakelite. Globe has been replaced. 40" H x 13" W (plate) x 10.5" D (plate), 11" D (base). $595-700.

1930s art deco airplane lighted smoking stand. Composition: white metal, chrome, and nickel-plated. Signed "Canadian Craft Mfg. Co." 34" H x 12.5" W (plate) x 9.5" D (plate), 10" D (base). $725-1100.

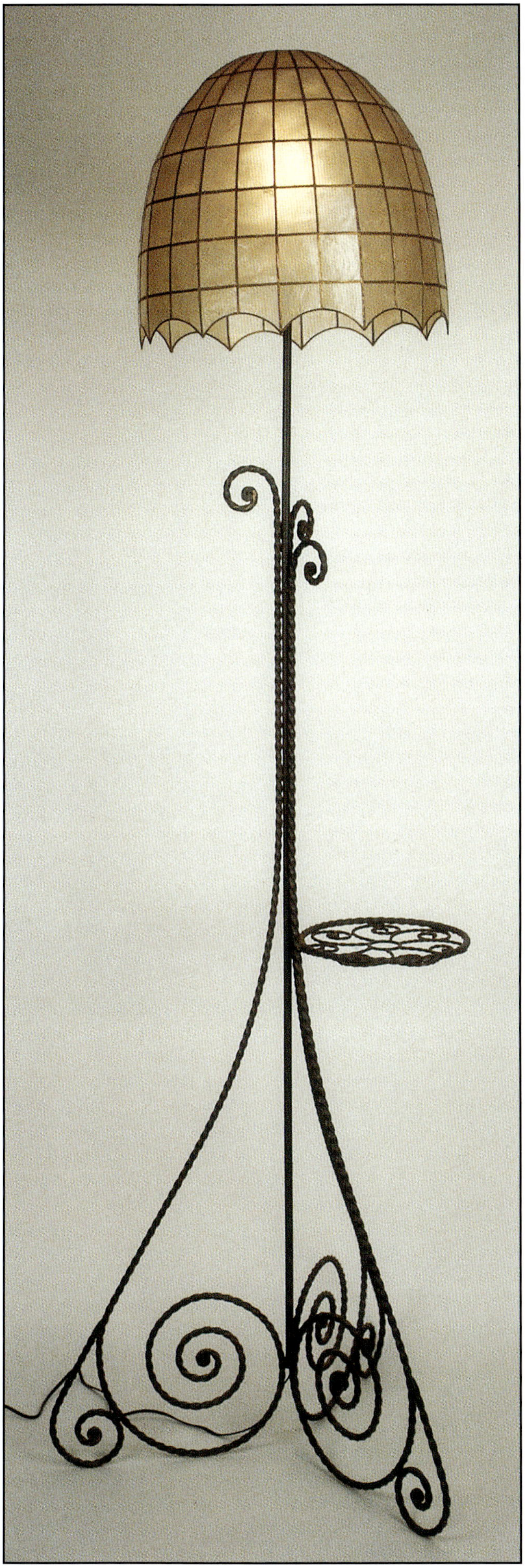

1930s French art deco floor lamp with side table. Composition: wrought iron, cappa shell shade has been replaced. 75" H x 24" W (base); 9" D (side table). $650-775.

1930s art deco bridge lamp. Composition: white metal and cast iron with pewter wash. Ornamental top and detail base has art nouveau influence. Period shade. 62" H x 11" W x 10" D. $625-695.

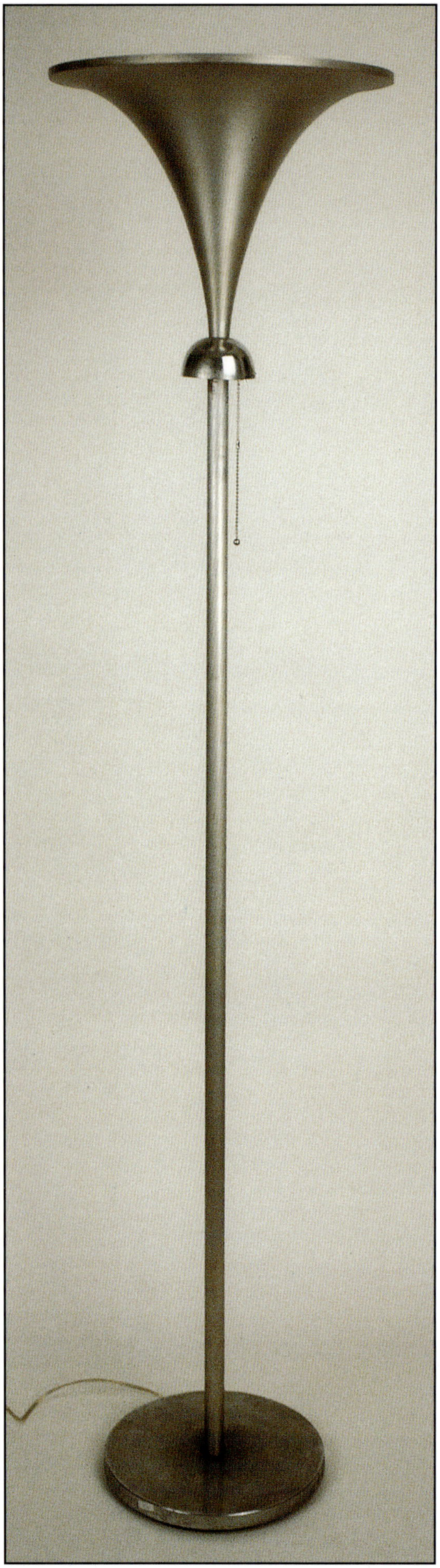

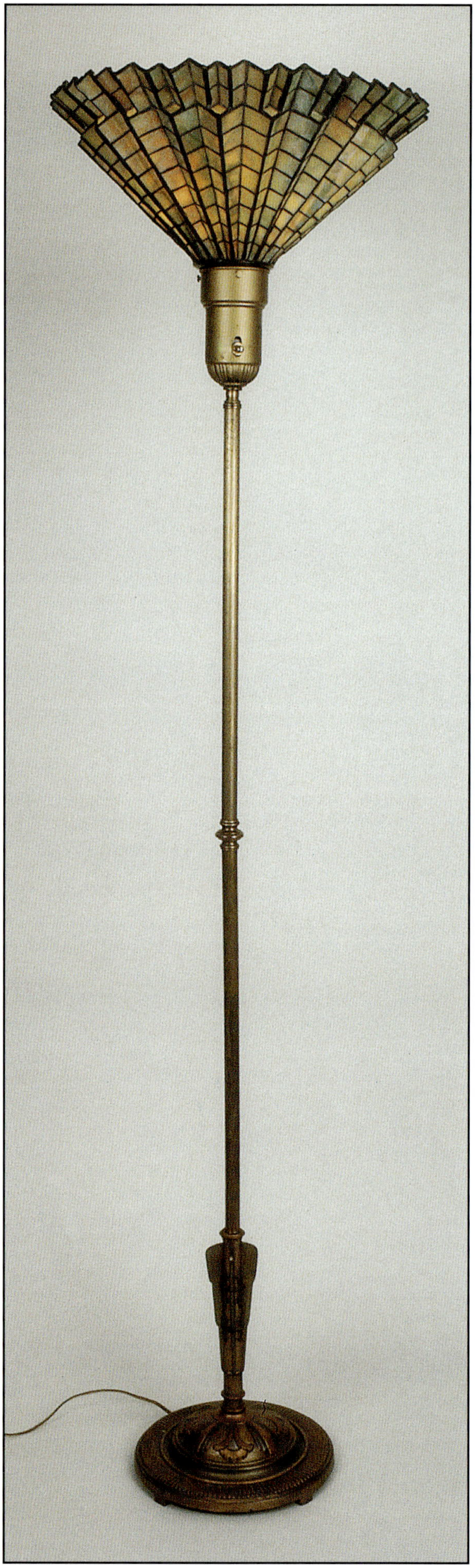

1930s art deco torchiere with flaring trumpet shade. Composition: brushed aluminum with chrome accents. 64.75" H x 11" D (base). $895-975.

1930s art deco torchiere. Period base with custom-made leaded glass shade. Composition: cast metal and brass, restored to its original finish. 63.5" H x 9.5" D (base); 16.5" D (shade). $1395-1500.

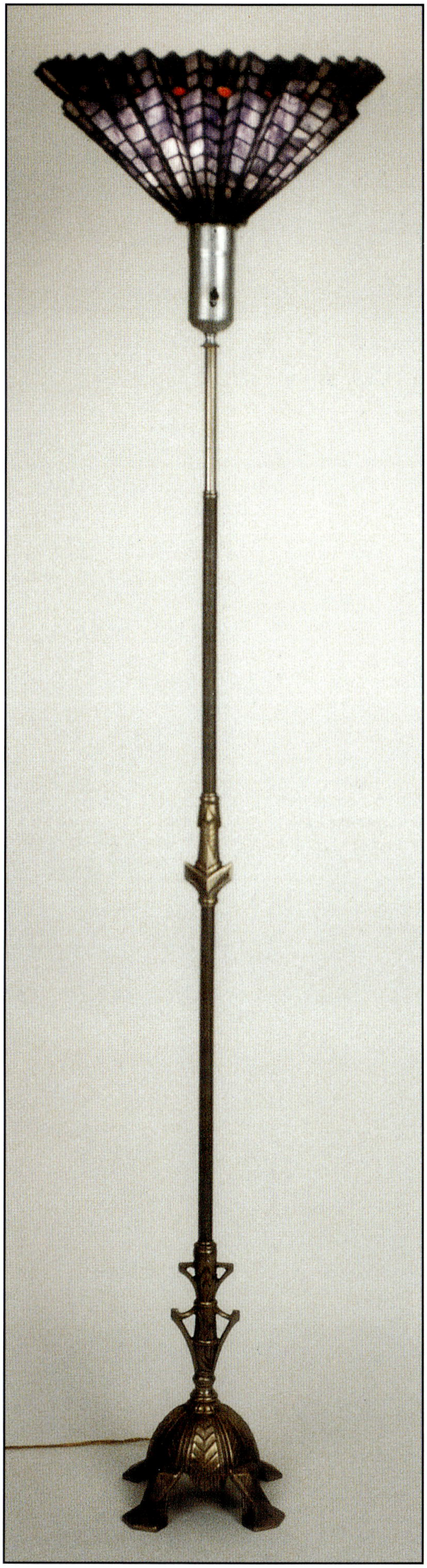

Late 1920s art deco torchiere. Custom leaded glass shade. Composition: cast iron, white metal with pewter and brass wash. 69.5" H x 8.25" W x 8.25" D. $1365-1550.

1930s art deco bridge lamp. Composition: white metal, chrome-plated, wood accents, with period shade. 54.5" H x 10" D (base). $515-625.

1930s art deco brass torchieres. 64" H x 10.5" D (base), 11.5" D (shade). $975-1100 pair.

Early 1930s art deco electric candelabras with original candle bulbs. Composition: brass and cast iron with original patina. 66.5" H x 10.5" D (base). $1695-1850 pair.

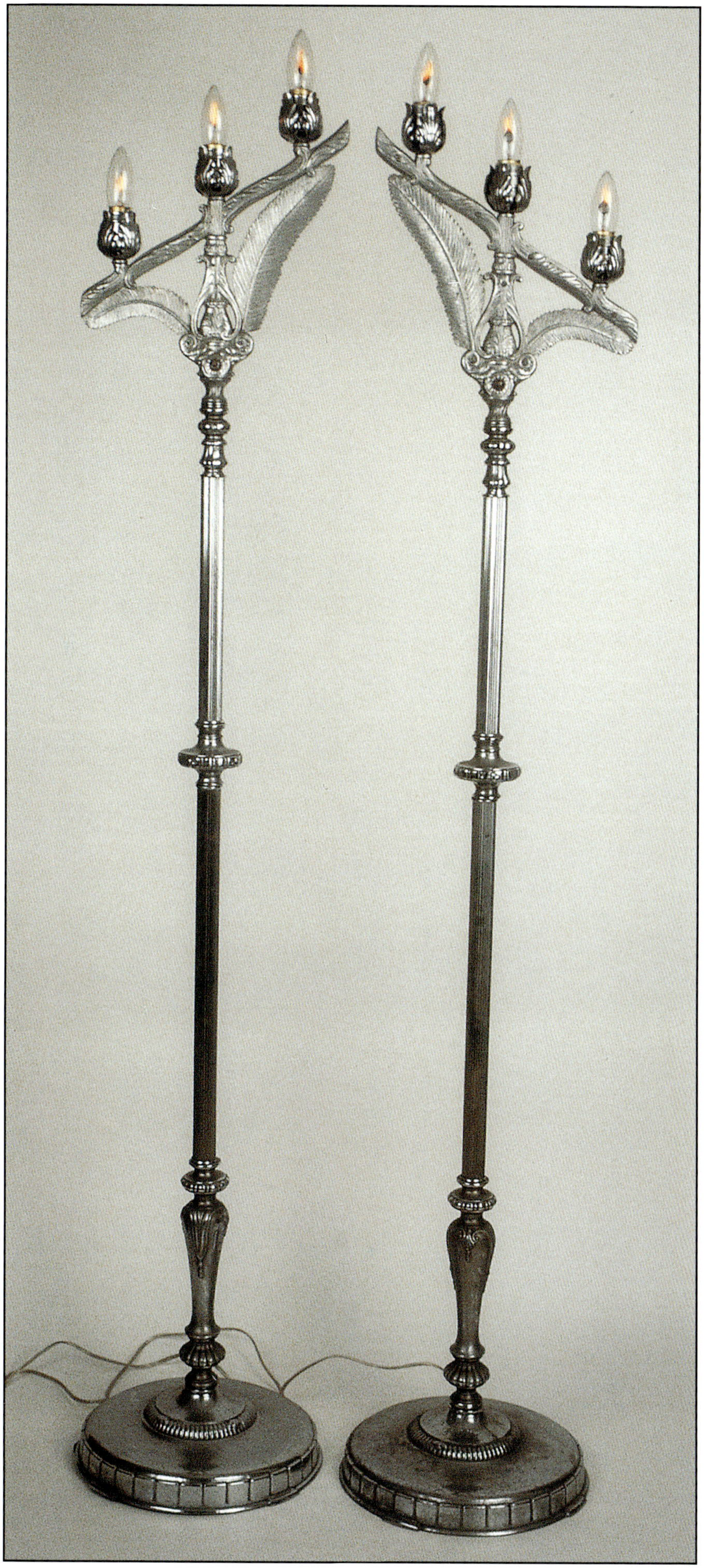

Early 1930s art deco electric floor candelabras. Composition: cast iron and aluminum, chrome-plated with pewter wash. 64.75" H x 12" D (base), $1365-1550 pair.

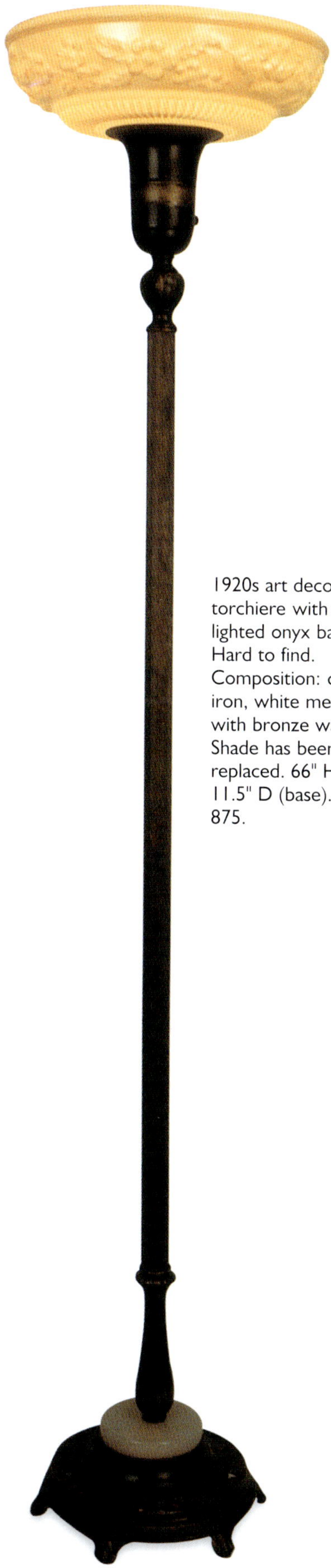

1920s art deco torchiere with lighted onyx base. Hard to find. Composition: cast iron, white metal with bronze wash. Shade has been replaced. 66" H x 11.5" D (base). $795-875.

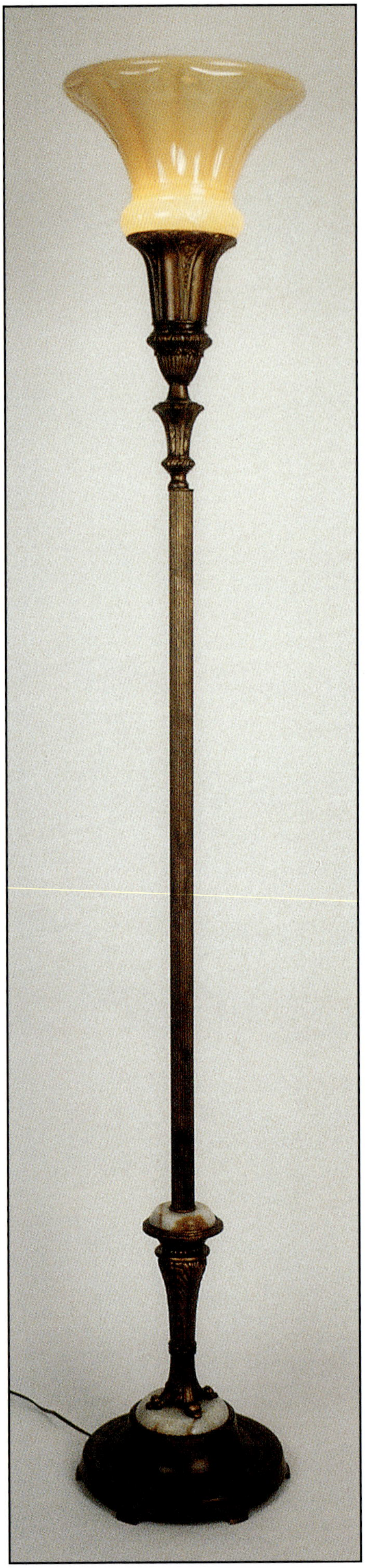

1930s art deco torchiere. Composition: cast iron, white metal with bronze wash. Shade has been replaced. 66.75" H x 11" D (base). $725-795.

Late 1920s art deco torchiere with two-tier onyx base. Composition: cast iron and white metal with bronze wash. Shade has been replaced. 70.75" H x 10" D (base). $825-945.

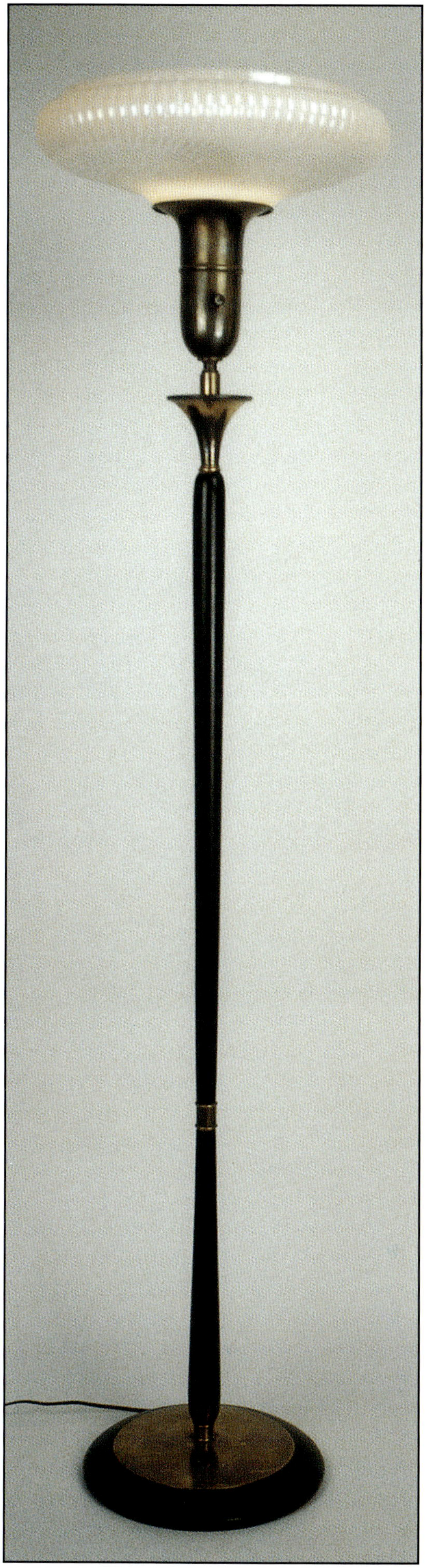

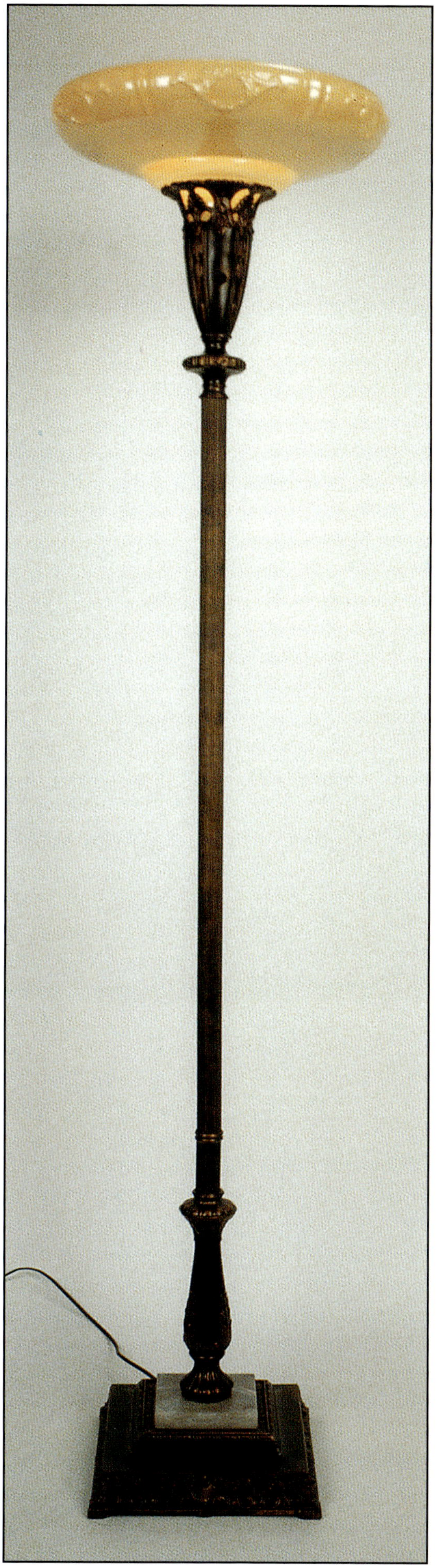

Early 1940s art deco torchiere. Composition: brass and black enameled wood. Signed "Rembrandt." Shade has been replaced. 64.75" H x 12" D (base). $565-650.

1930s art deco torchiere with square onyx base. Very ornate. Composition: cast iron and white metal with bronze wash. Shade has been replaced. 65.25" H x 11.25" W x 11.25" D. $895-965.

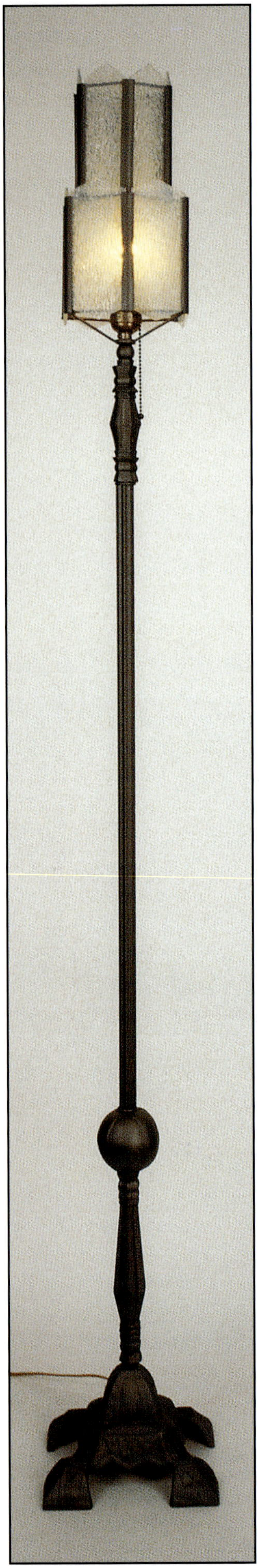

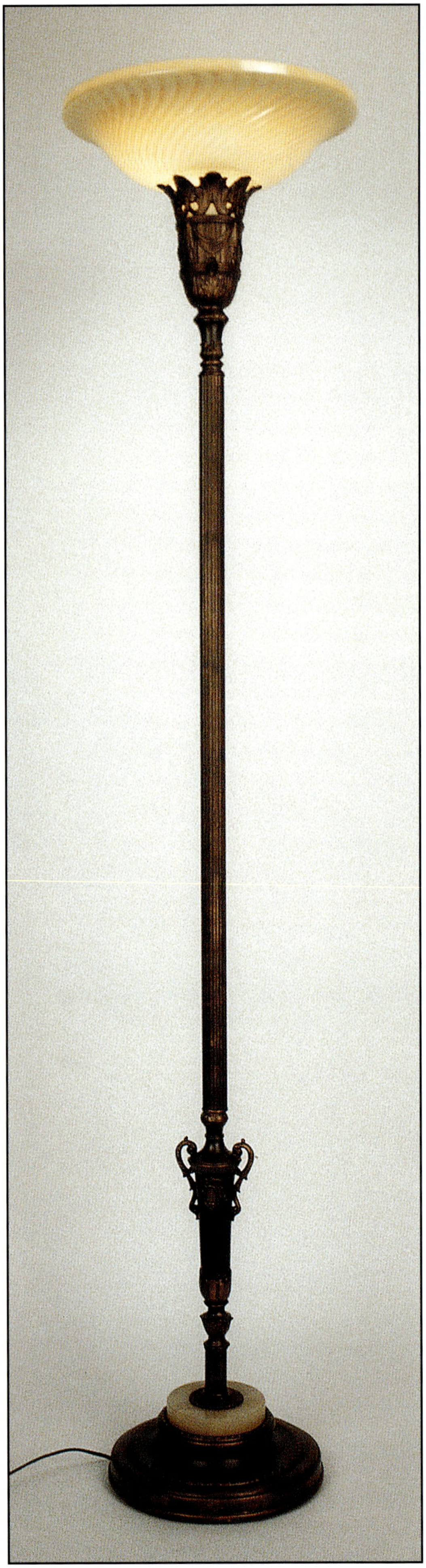

1930s art deco floor lamp. Composition: cast iron and white metal with pewter wash. Twelve panels of iced glass have been replaced. 65.75" H x 8.75" W x 8.75" D. $595-685.

1930s art deco torchiere with circular onyx base. Composition: cast iron, white metal with bronze wash. Shade has been replaced. 66" H x 9.5" D (base). $845-895.

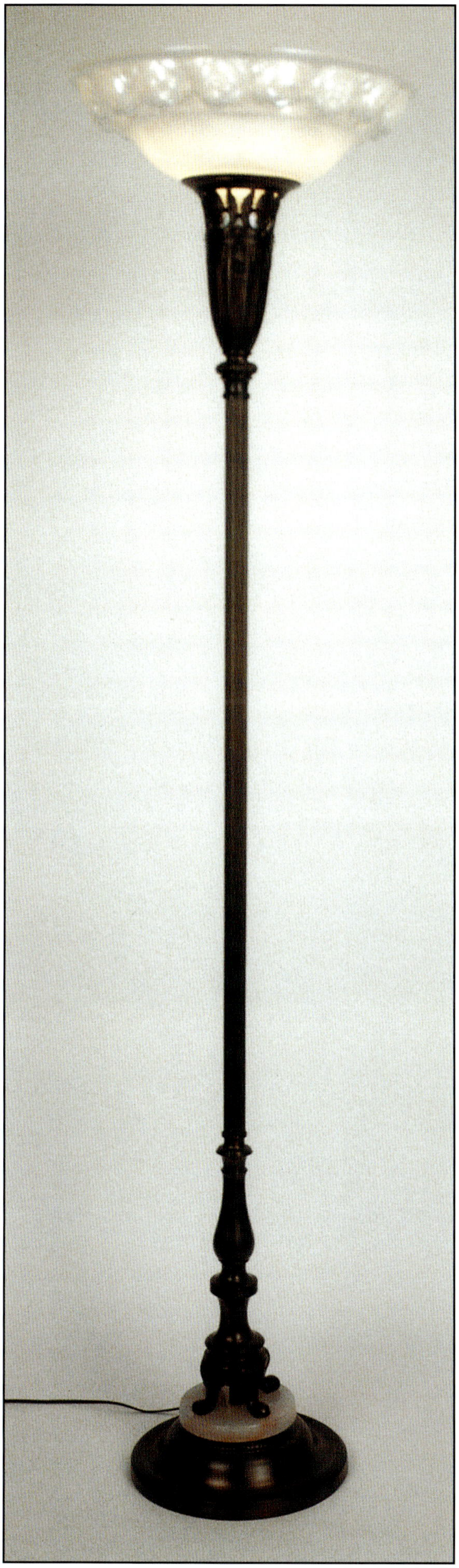

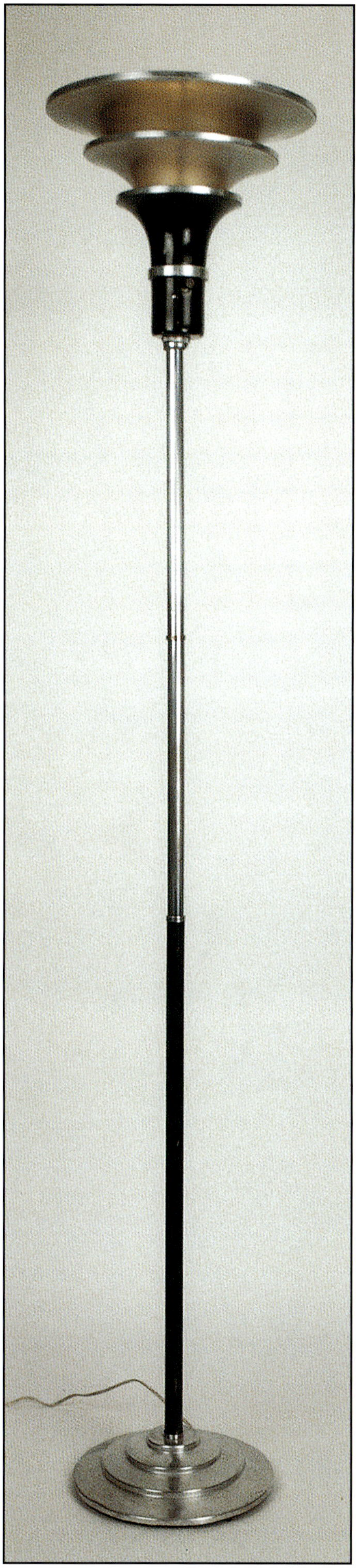

1930s art deco torchiere with circular onyx base. Composition: white metal and cast iron with bronze wash. Shade has been replaced. 65.75" H x 10.75" D (base). $825-910.

1930s art deco torchiere, triple tier with mogul light base. Composition: aluminum and chrome with black enamel highlights. 64.5" H x 9.75" D (base). $895-975.

Late 1920s art deco mini torchieres. Strong Grecian detail with original cranberry glass inserts. Composition: cast iron and white metal with gold and black wash. 50" H x 10.75" W x 10.75" D. $1895-2100 pair.

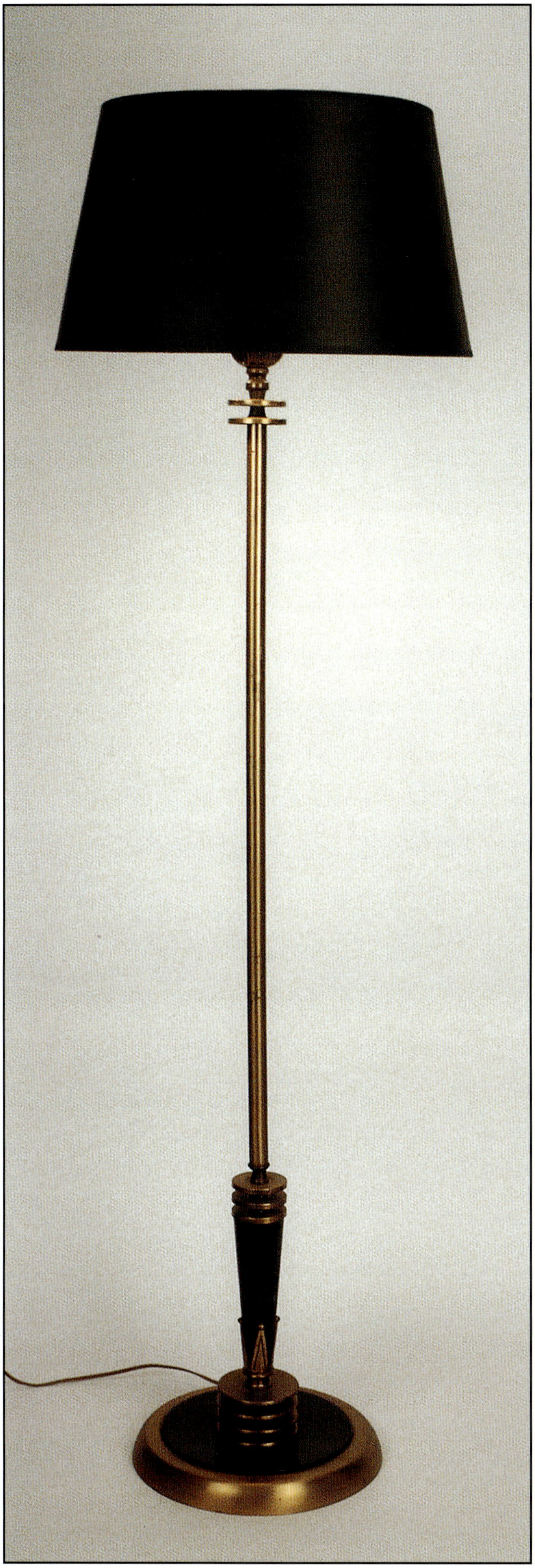

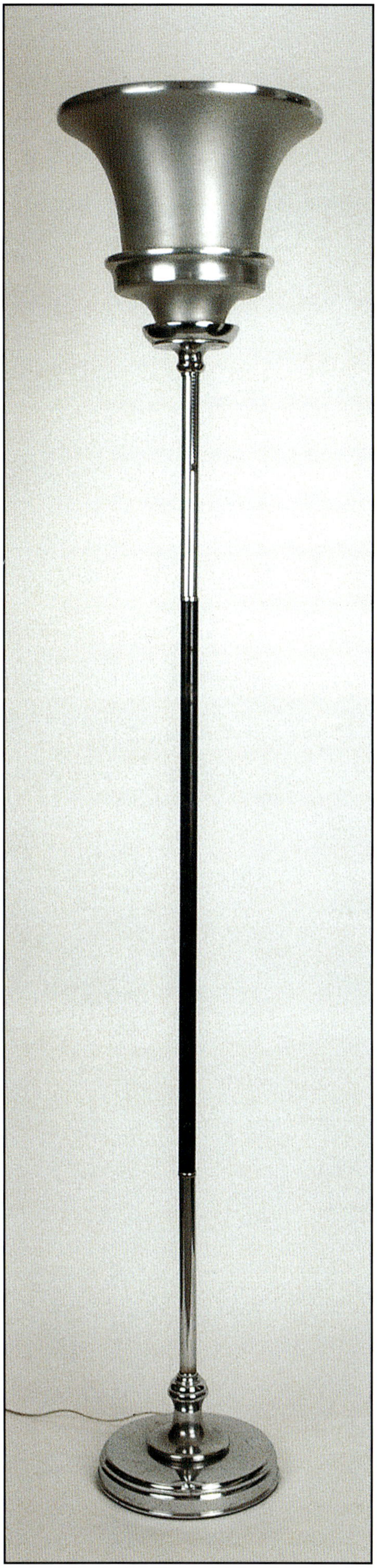

1930s art deco floor lamp. Composition: brass and black enamel. Shade has been replaced. Base is machine age in design. 60" H x 12" D (base). $895-1050.

1930s art deco torchiere. Composition: aluminum and chrome with black enamel accents. 64.5" H x 9" D (base). $895-975.

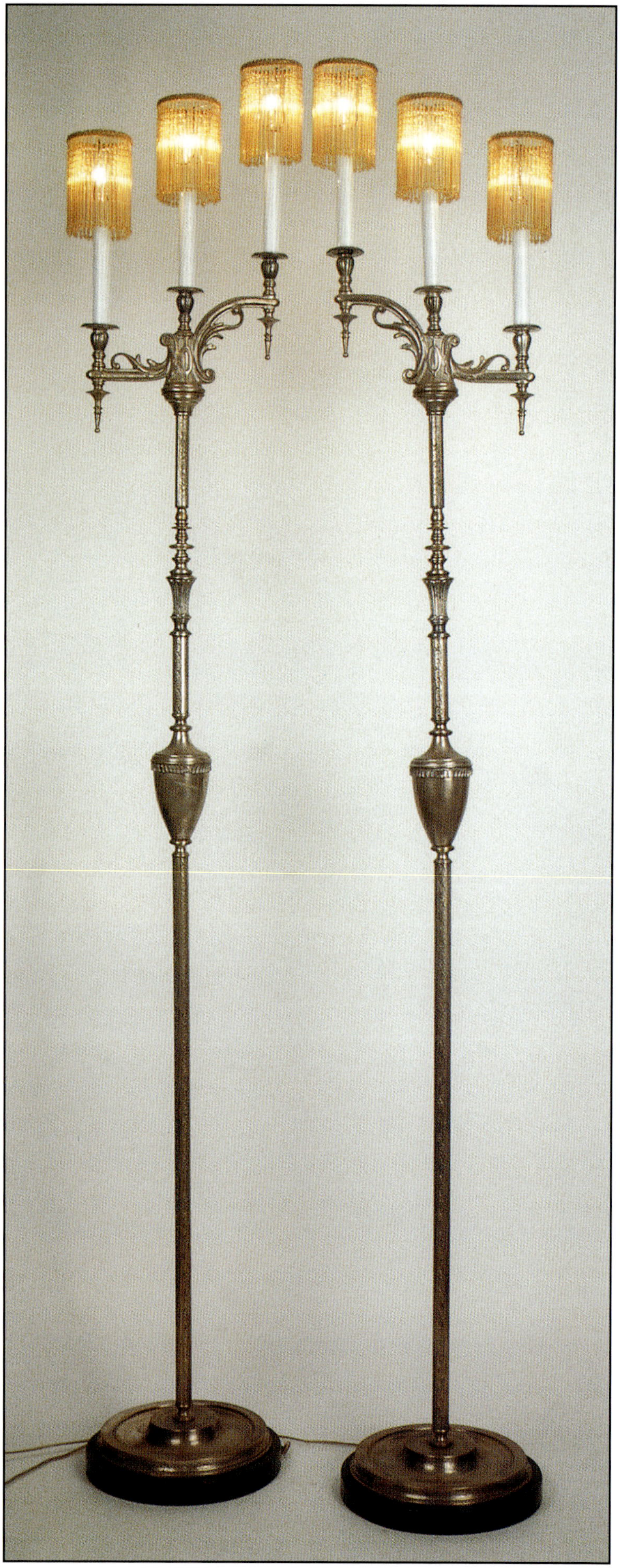

1910-1915 art deco banquet lamps. Composition: silver-plated, custom glass beaded shades. From an estate in South Hampton, Long Island, New York. 71" H x 10.5" D (base). $2800-3200 pair.

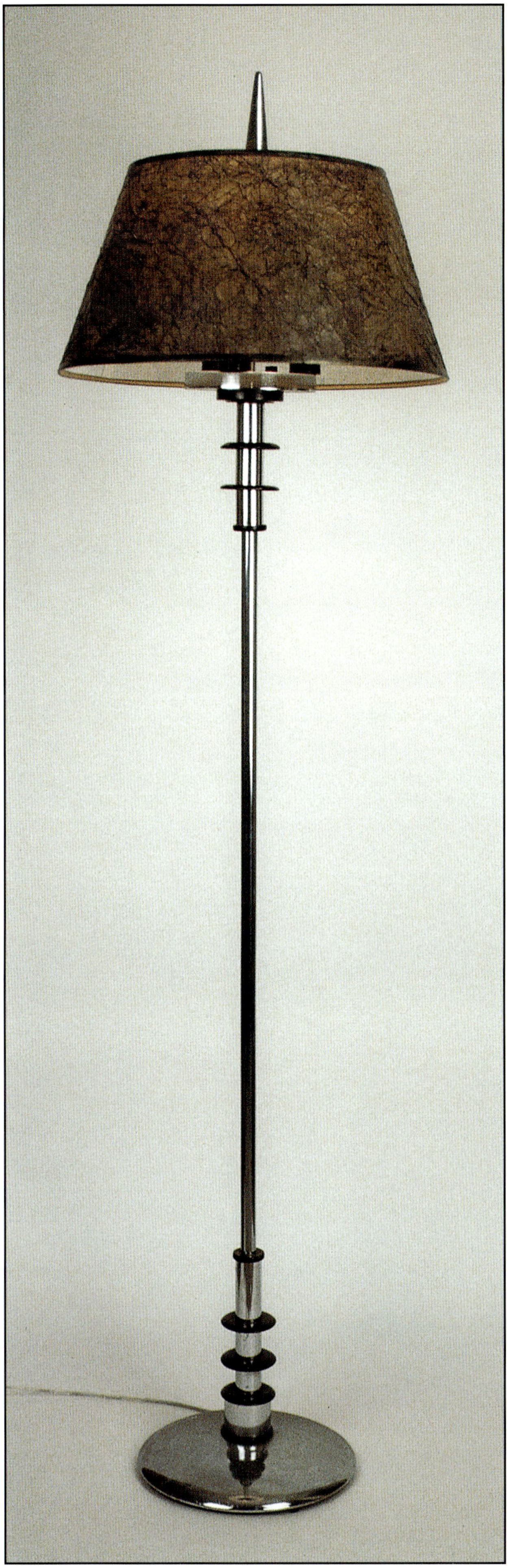

1930s art deco floor lamp. Great lines. Composition: aluminum and chrome with black Bakelite rings. Shade has been replaced. 62.5" H x 10" D (base). $895-975.

Early 1930s art deco floor lamp. Composition: solid brass. Very unusual star base. Silk shade is a replacement. 64.75" H x 13.25" D (base). $995-1125.

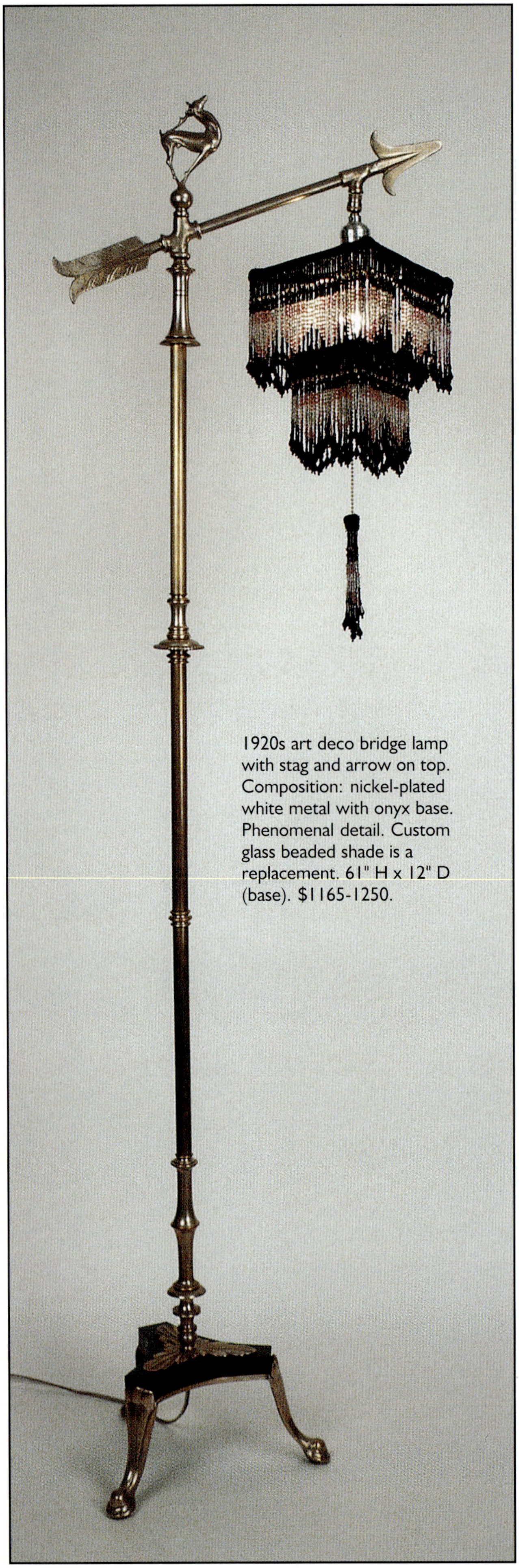

1920s art deco bridge lamp with stag and arrow on top. Composition: nickel-plated white metal with onyx base. Phenomenal detail. Custom glass beaded shade is a replacement. 61" H x 12" D (base). $1165-1250.

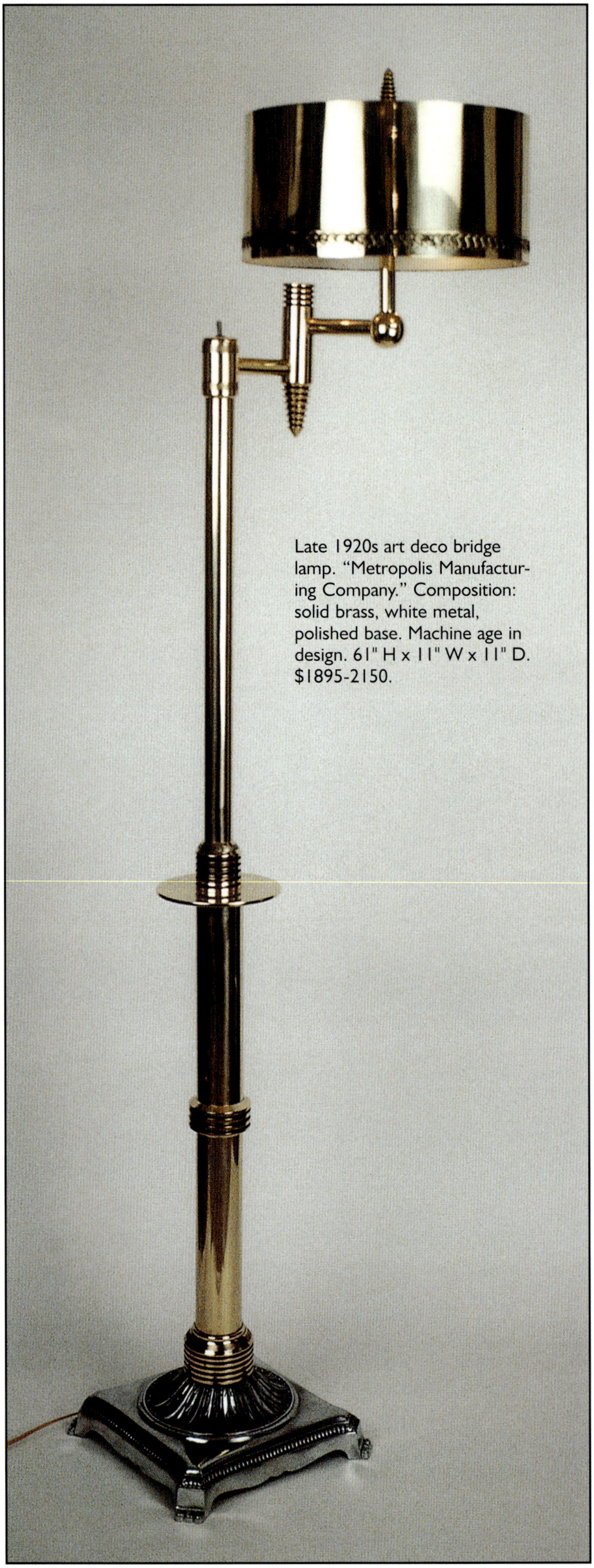

Late 1920s art deco bridge lamp. "Metropolis Manufacturing Company." Composition: solid brass, white metal, polished base. Machine age in design. 61" H x 11" W x 11" D. $1895-2150.

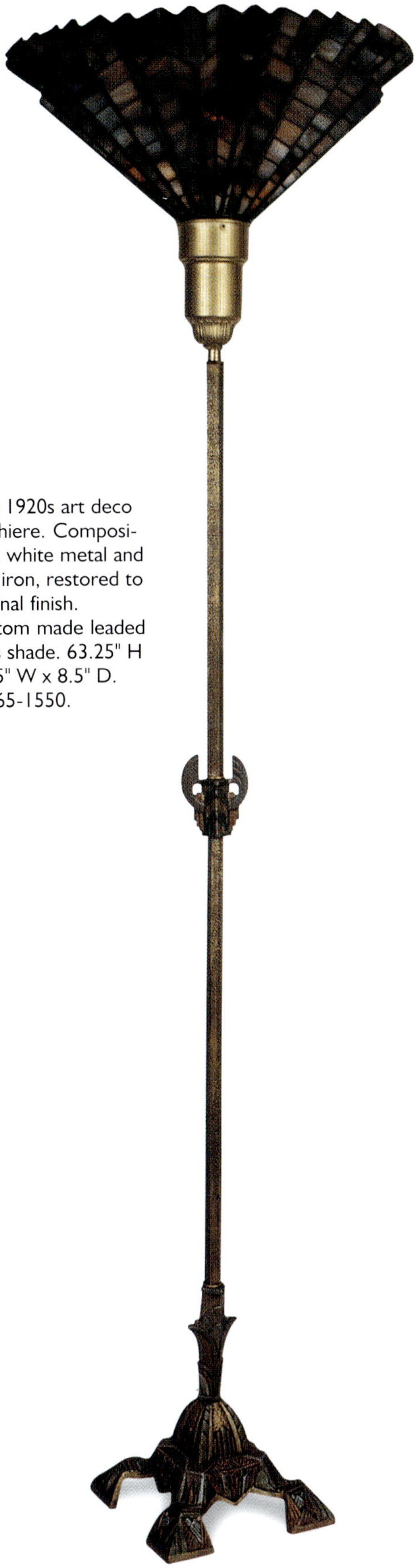

Late 1920s art deco torchiere. Composition: white metal and cast iron, restored to original finish. Custom made leaded glass shade. 63.25" H x 8.5" W x 8.5" D. $1365-1550.

Late 1920s art deco floor lamp. Composition: white metal and cast iron. Very unusual base with three female nudes. Shade and finial have been replaced. 52.75" H x 8.250" D (base). $985-1155.

Early 1930s art deco torchiere with circular onyx base. Composition: cast iron and white metal with bronze wash. Opalescent glass shade has been replaced. 55.5" H x 11.75" D (base). $675-750.

Early 1940s art deco bridge lamp. Composition: cast iron, white metal, and Bakelite, chrome-plated with black enamel accents. Paper shade has been replaced. 57.75" H x 11" D (base). $345-465.

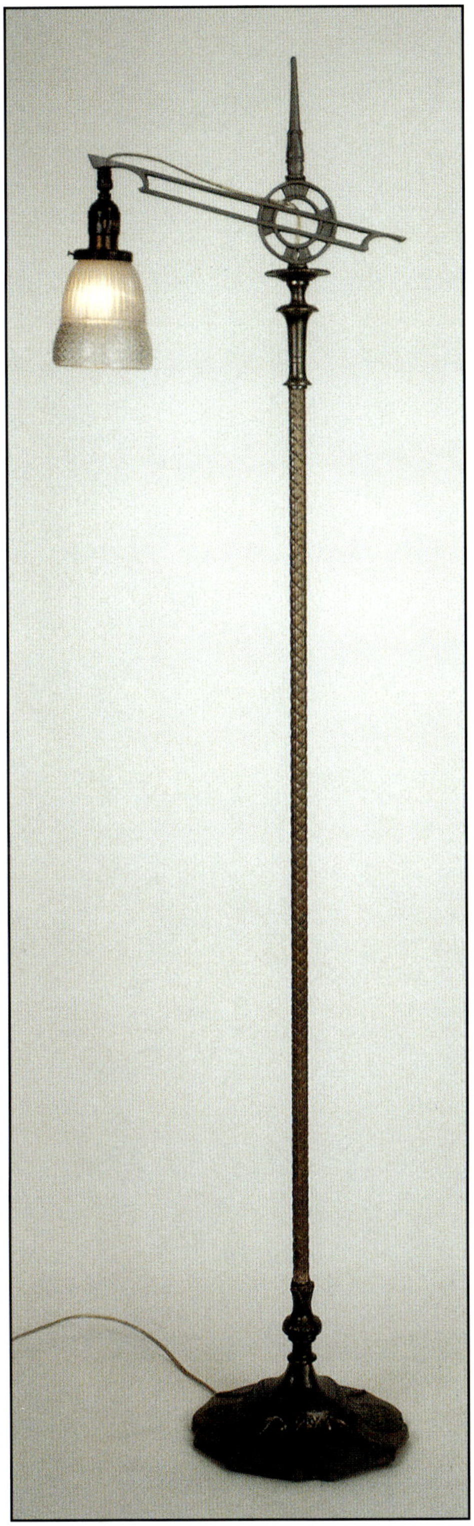

1930s art deco bridge lamp. Composition: cast iron and white metal. Industrial design bridge with period shade, pressed glass. 58.5" H x 10" D (base). $465-545.

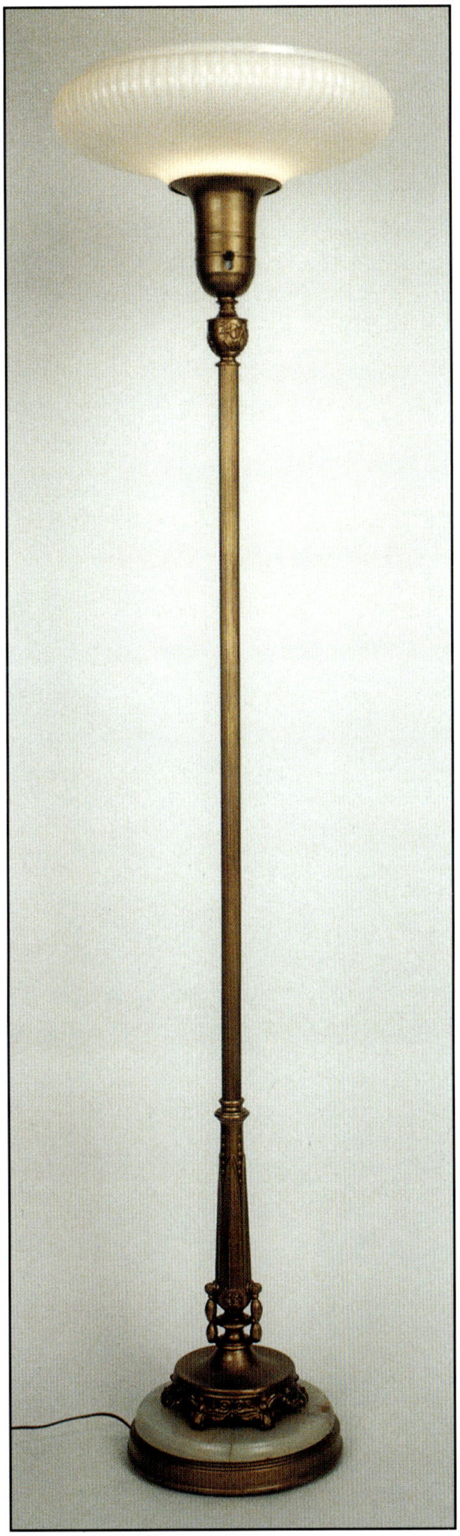

1930s art deco torchiere with circular onyx base. Composition: cast iron and white metal with gold tone wash. Opalescent glass shade has been replaced. 66" H x 11" D (base). $795-865.

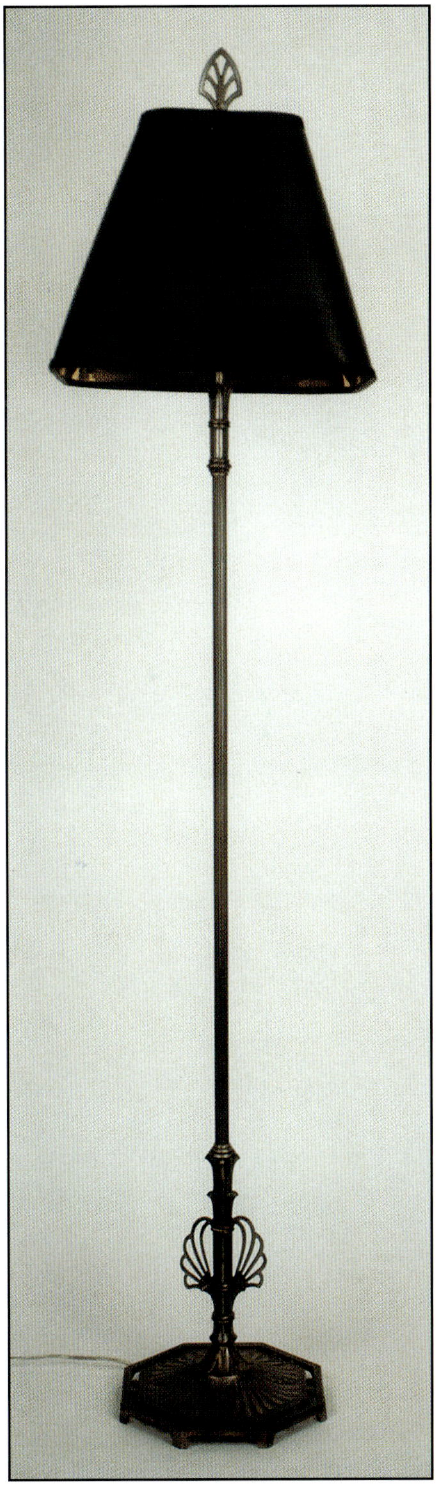

1920s art deco floor lamp. Signed "Hubley." Composition: cast iron and white metal with pewter wash. Paper shade is a replacement. 61.25" H x 10" D (base). $685-765.

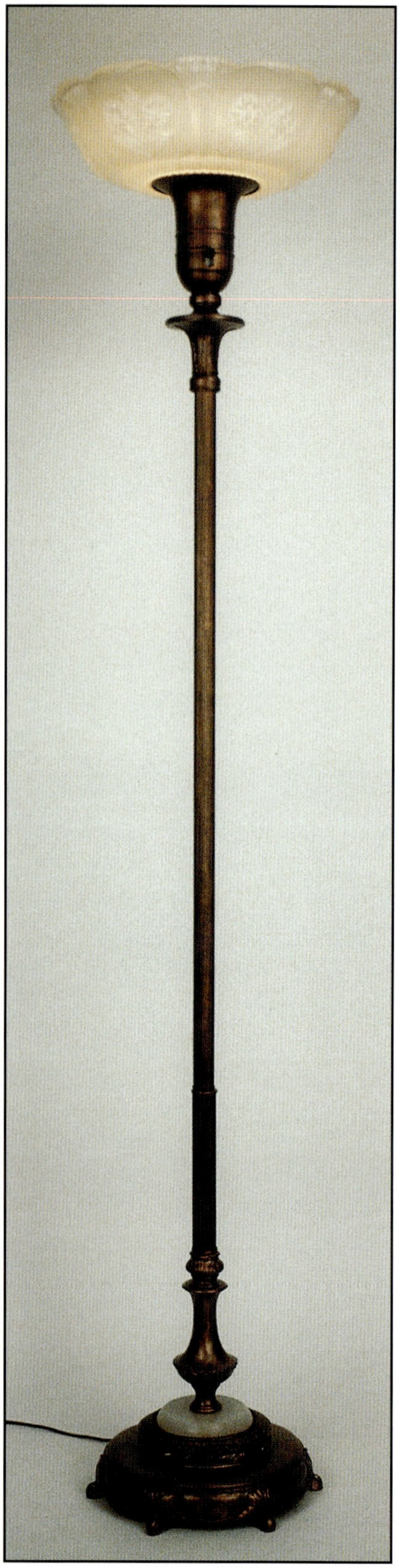

1920s art deco torchiere with circular onyx base. Original Phoenix glass shade. Composition: cast iron and white metal with bronze wash. 68.25" H x 11.5" D (base). $895-975.

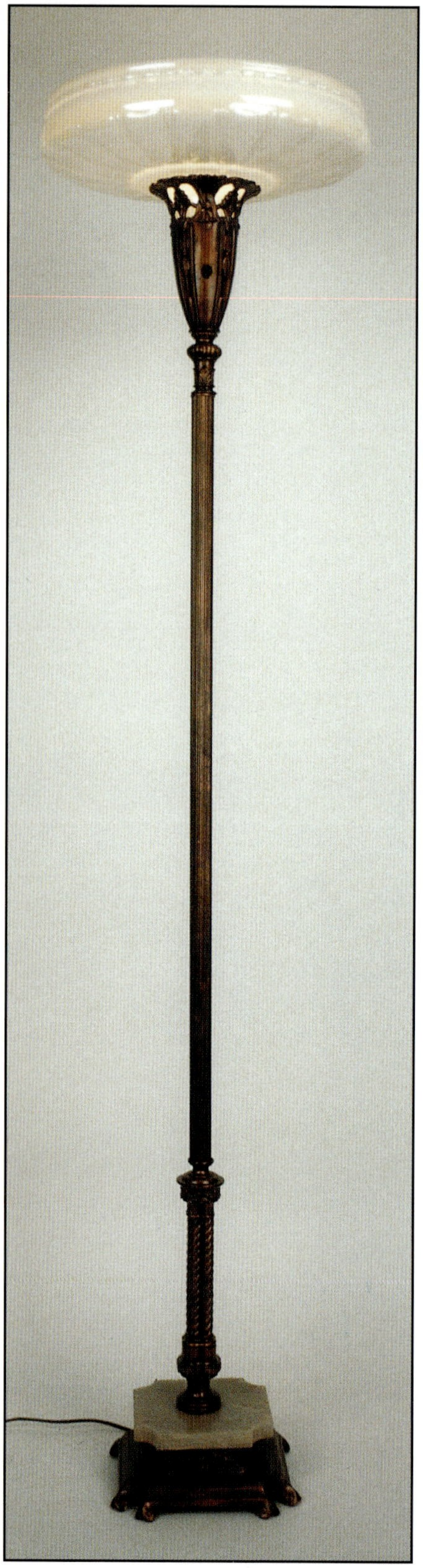

1930s art deco torchiere with circular onyx base. Composition: cast iron and white metal with bronze wash. Opalescent flowered glass shade has been replaced. 66" H x 11" D (base). $825-910.

Late 1920s art deco torchiere, "Metropolis Manufacturing Company." A very rare find. Industrial in design with downlights, as well as mogul globe uplight. Separate switches control lower and upper lights. Composition: solid brass. 64.75" H x 12" D (base). $1999-2350.

Early 1930s art deco bridge lamp and smoking stand with period hallophane shade. Composition: cast iron, nickel, and chrome-plated with black onyx accents. 57.25" H x 10.5" D (base) x 12.25" W (smoking stand). $915-1075.

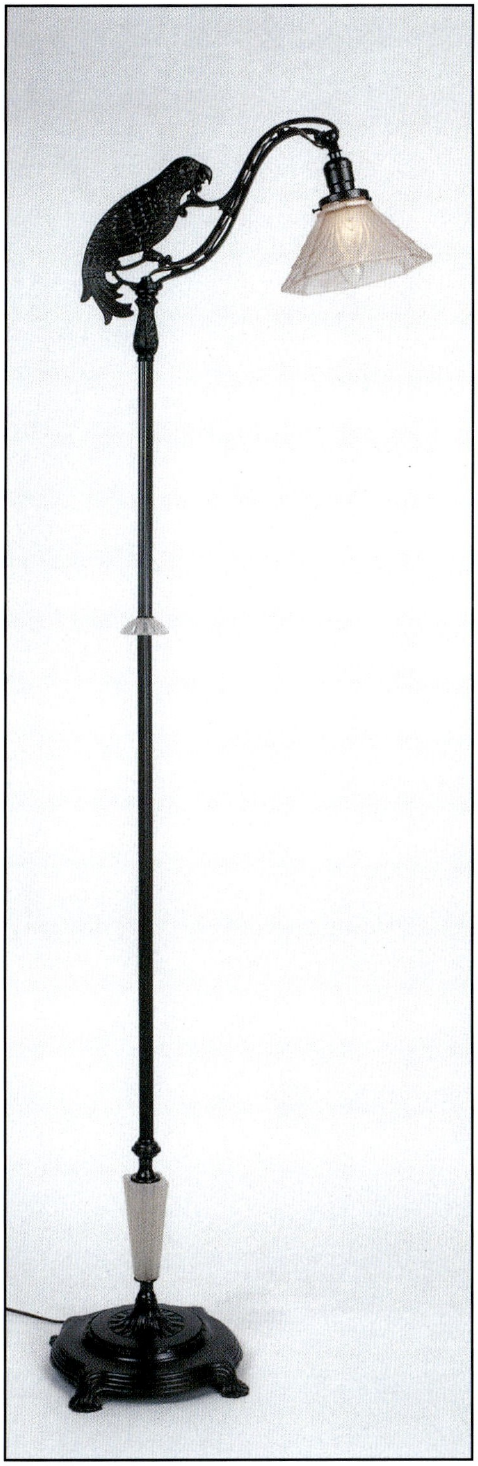

1920s art deco parrot bridge lamp. Composition: white metal and cast iron with period pink French satin glass shade and pink slag glass accents on the shaft of the lamp. 61.5" H x 12" D (base). $685-795.

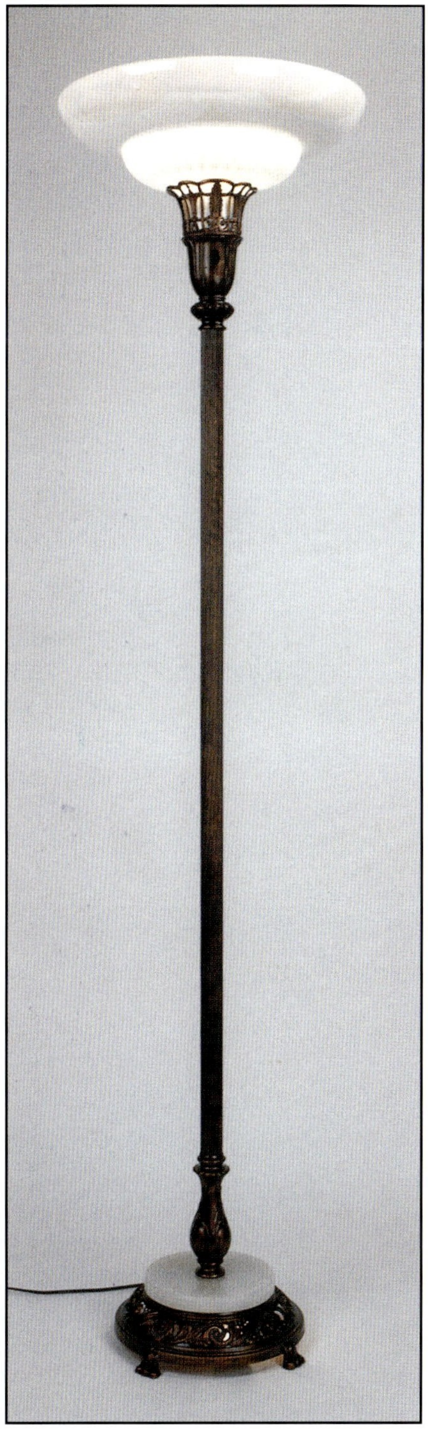

1920s art deco torchiere with onyx lighted base. A rare find. Composition: white cast metal with bronze wash. Shade has been replaced 65" H x 12.5" D (base). $865-945.

Bibliography

Thomas, Jo Ann. *Lighting Fixtures of the Depression, Book I.* Paducah, Kentucky: Collector Books-Schroeder Publishing, 2001.

Thomas, Jo Ann. *Lighting Fixtures of the Depression, Book II.* Paducah, Kentucky: Collector Books-Schroeder Publishing, 2001.